Lecture Notes in Physics

New Series m: Monographs

The Editorial Policy for Monographs

The series Lecture Notes in Physics reports new developments in physical research and teaching - quickly, informally, and at a high level. The type of material considered for publication in the New Series m includes monographs presenting original research or new angles in a classical field. The timeliness of a manuscript is more important than its form, which may be preliminary or tentative. Manuscripts should be reasonably self-contained. They will often present not only results of the author(s) but also related work by other people and will provide sufficient motivation, examples, and applications.
The manuscripts or a detailed description thereof should be submitted either to one of the series editors or to the managing editor. The proposal is then carefully refereed. A final decision concerning publication can often only be made on the basis of the complete manuscript, but otherwise the editors will try to make a preliminary decision as definite as they can on the basis of the available information.
Manuscripts should be no less than 100 and preferably no more than 400 pages in length. Final manuscripts should preferably be in English, or possibly in French or German. They should include a table of contents and an informative introduction accessible also to readers not particularly familiar with the topic treated. Authors are free to use the material in other publications. However, if extensive use is made elsewhere, the publisher should be informed. Authors receive jointly 50 complimentary copies of their book. They are entitled to purchase further copies of their book at a reduced rate. As a rule no reprints of individual contributions can be supplied. No royalty is paid on Lecture Notes in Physics volumes. Commitment to publish is made by letter of interest rather than by signing a formal contract. Springer-Verlag secures the copyright for each volume.

The Production Process

The books are hardbound, and quality paper appropriate to the needs of the author(s) is used. Publication time is about ten weeks. More than twenty years of experience guarantee authors the best possible service. To reach the goal of rapid publication at a low price the technique of photographic reproduction from a camera-ready manuscript was chosen. This process shifts the main responsibility for the technical quality considerably from the publisher to the author. We therefore urge all authors to observe very carefully our guidelines for the preparation of camera-ready manuscripts, which we will supply on request. This applies especially to the quality of figures and halftones submitted for publication. Figures should be submitted as originals or glossy prints, as very often Xerox copies are not suitable for reproduction. For the same reason, any writing within figures should not be smaller than 2.5 mm. It might be useful to look at some of the volumes already published or, especially if some atypical text is planned, to write to the Physics Editorial Department of Springer-Verlag direct. This avoids mistakes and time-consuming correspondence during the production period.
As a special service, we offer free of charge LaTeX and TeX macro packages to format the text according to Springer-Verlag's quality requirements. We strongly recommend authors to make use of this offer, as the result will be a book of considerably improved technical quality.
Manuscripts not meeting the technical standard of the series will have to be returned for improvement.
For further information please contact Springer-Verlag, Physics Editorial Department II, Tiergartenstrasse 17, D-69121 Heidelberg, FRG.

A.V. Bogdanov G.V. Dubrovskiy M.P. Krutikov
D.V. Kulginov V.M. Strelchenya

Interaction of Gases with Surfaces

Detailed Description of Elementary Processes and Kinetics

Springer

Authors

Alexander V. Bogdanov
German V. Dubrovskiy
Michael P. Krutikov
Dmitry V. Kulginov
Institute for Interphase Interactions
P. O. Box 1146
St. Petersburg 194291, Russia

Victor M. Strelchenya
Institute for Interphase Interactions
P. O. Box 84
St. Petersburg 197373, Russia

ISBN 978-3-662-14042-0 ISBN 978-3-540-49107-1 (eBook)
DOI 10.1007/978-3-540-49107-1

CIP data applied for.

Originally published by Springer-Verlag Berlin Heidelberg New York in 1995
Softcover reprint of the hardcover 1st edition 1995

This book was processed by the authors using the TEX macro package from Springer-Verlag Berlin Heidelberg GmbH.

SPIN: 10080298 55/3140-543210 - Printed on acid-free paper

Dim vales – and shadowy floods ·
And cloudy-looking woods,
Whose forms we can't discover
For the tears that drop all over...

Edgar Allan Poe

Dedication

To our women, whose endurance and goodwill brought this book to its present state.

Foreword

This book is one of the first attempts to give a unified description of surface processes, i.e. elementary events, surface kinetics, surface influence upon gas flow etc. Such a description proved to be possible in the framework of the quasiclassical approach proposed by the authors and turned out to be useful in numerous applications. In the consideration of the physical picture of gas–surface scattering we paid the greatest attention to understanding basic laws and mechanisms of phenomena, so we limited the discussion to only the most essential experimental results.

This book can be recommended as a textbook on gas–surface interface phenomena, their interpretation, and the methods of theoretical understanding of the pertinent physical laws.

Acknowledgements

This book was written during the period 1990–1993. There have been several versions of it, and even the list of authors changed several times. We are not sure that this version is the best one. Nevertheless it is the final one and readers must judge it for themselves. It is our pleasure to thank all our colleagues who made a contribution to the book. First of all, we thank the members of our seminar on gas–surface interaction at the Institute of Interphase Interaction, State Technical University, St. Petersburg, Russia. We appreciate very much all our guests, who kept us up-to-date on recent research and provided fruitful discussions of our methods and results. Our special acknowlegements go to Drs. Nicholas V. Blinov, Vladimir A. Fedotov, Yuri E. Gorbachev, Yuri G. Markoff, and Dmitry A. Shapiro.

Preface

Interface phenomena are the most fascinating and interesting to a researcher because of the mixing of different scales and interference of physical processes of diverse nature. Even if the understanding of the processes in separate phases can be regarded as more or less clear (although a lot of researchers would challenge this assumption), bringing them together produces a mixture of competing tendencies and their non-linear interference, so that even qualitative consideration of the interface phenomena presents serious difficulties.

The fact that many different scales play a role in gas–surface scattering makes it necessary to use different levels of description: microscopic, kinetic, and gas-dynamical. Each of these levels is treated in its own way and specific features of each particular level bring specific tools for studying gas–surface interaction phenomena. In practice this means that for the interpretation of some particular effects certain models, not properly matched among themselves, are used on each level of description.

To answer practical questions dealing with inelastic gas–surface scattering, kinetics of adsorption layers, the evolution of inhomogeneities and defects at the surface, the account of a Knudsen layer, the development of boundary conditions on the kinetic and gas-dynamical levels, the determination of exchange and slip coefficients, etc. one needs to use a unified approach. Such an approach should provide a consistent accuracy of the description at different levels and make it possible to deduce the observables on one level from those of another one. This problem is rather difficult because of the difference in the methods of consideration and in the physical nature of observables on every level.

We hope that such problems might be solved by the application of quasiclassical methodology. Despite the difference in the nature of physical phenomena described, the quasiclassical results are internally self-consistent and may be applied for the investigation of scattering phenomena at real surfaces. Such phenomena are important for aerodynamics and gas-dynamics of hypersonic flows, aerothermochemistry of flying vehicles, thin film growing technologies, biology and ecology.

It is inevitable that the discussion of such serious problems is overwhelmed with a lot of details. That is why we try to restrict ourselves to a comprehensive discussion of the formulation of a problem and propose some effective means for its solution. Although attention is paid mainly to the physical ideas and problems of compatibility of different approaches important for the providing of theoretical background for our methodology, sometimes (where it is relevant) we give a comparison with experimental data.

The book consists of four main parts. In the first one we discuss the questions of molecule scattering. In the second part some questions of adsorption kinetics are approached. The third part deals with some phenomenological models of thin film growth. And in the last part we discuss the kinetic boundary conditions and inverse scattering problem.

Formally speaking the microscopic results given in Part I are too detailed to be used in the kinetic models discussed in Parts II and III. But on the basis of these results more delicate problems of kinetics may be approached using the same methodology (for example, the problems dealing with dynamical models of surface roughening due to adsorption layers and its influence upon gas–surface scattering).

The system of units with $\hbar = 1$ is used throughout the book.

Contents

I

Inelastic Scattering of Molecules from Crystal Surfaces

Introduction

The semiclassical approach to the problem of atom–crystal inelastic scattering is very attractive due to its relative simplicity, analytical nature and wide applicability. This approach allows one to obtain a simple Gaussian approximation (Brako and Newns 1982; Manson 1991) to the dynamic structural factor of inelastic phonon scattering and the intensities of diffraction peaks (Billing 1975). The effect of umklapp processes on the dynamic structural factor has been considered only in the hard-wall approximation (Berry 1975; Bogdanov 1980) or numerically (Manson 1991).

There are two main quasiclassical approaches to the gas–surface scattering problem. One of them (Bogdanov 1980) using the generalized eikonal method developed in molecular collision theory (Dubrovskiy and Bogdanov 1979a; Bogdanov et al. 1989) is based on path integration over projectile variables and allows one to describe inelasticity and diffraction, but it treats the surface classically and does not take into account the softness of the potential and tangential displacements of the crystal atoms.

Another approach (Newns 1985; Nourtier 1985) is based on the double path integral representation of the scattering probability. The surface dynamics is described in terms of an influence functional that can be found in explicit form in the linear coupling approximation. However, semiclassical evaluation of the path integrals leads to a complex classical trajectory problem that is non-local in time and too difficult for applications.

A formalism presented in this part unifies both approaches. Path integrals over projectile variables in momentum representation are evaluated in the quasiclassical limit separately before and after the turning point in the spirit of generalized eikonal method. The Faddeev–Popov method (Popov 1983) is used to fix classical trajectories with respect to the symmetry of the problem. The influence functional is treated as a pre-exponential factor.

The suggested formalism is a convenient tool for the development of further approximations corresponding to various physical situations.

1 General Semiclassical Theory of Gas–Surface Scattering

1.1 Gas–Surface Interaction Potential

Owing to the short-range nature of the interatomic pair potential, the non-stationary part of the atom–surface interaction is assumed to be mainly contributed by the first layer of surface atoms. The instant configuration of this layer is described by the displacement $\boldsymbol{u}_k$ of kth atom from its equilibrium position $\boldsymbol{R}_k$. It is convenient to express $\boldsymbol{u}_k$ through the amplitudes a_l of phonon modes $l = (\boldsymbol{q}, j)$ with wave vectors $\boldsymbol{q}$ and polarizations j (the system of units with $\hbar = 1$ is used throughout this review):

$$\boldsymbol{u}_k = \sum_l \frac{\boldsymbol{e}_l}{\sqrt{2\omega_l MN}} [\exp(\mathrm{i}\boldsymbol{q}\boldsymbol{R}_k) a_l + \exp(-\mathrm{i}\boldsymbol{q}\boldsymbol{R}_k) a_l^*], \tag{1.1.1}$$

here M is the mass of crystal atom, N is the number of atoms in the crystal block; dispersion ω_l and polarization vectors e_l are determined by the structure of both bulk and surface phonon modes.

Within the linear coupling approximation the interaction potential is

$$V(\boldsymbol{r},\{\boldsymbol{u}_k\}) = V_s(\boldsymbol{r}) - \sum_l (a_l^* f_l + a_l f_l^*), \tag{1.1.2}$$

where $\boldsymbol{r}$ is the radius vector of the gas atom (the axis z is normal to the surface, bold-face capitals refer to projections of vectors onto the surface) and f_m is the force of the interaction of the gas particle with phonon mode $m = (\boldsymbol{q}, j)$:

$$f_l = -\frac{\boldsymbol{e}_l}{\sqrt{2\omega_l MN}} \sum_k \exp(-\mathrm{i}\boldsymbol{q}\boldsymbol{R}_k)\frac{\partial V}{\partial \boldsymbol{u}_k}. \tag{1.1.3}$$

For the static atom–surface potential an expression

$$V_s(\boldsymbol{r}) = V_0(z) + \sum_{\boldsymbol{G}} V_{\boldsymbol{G}}(z)\exp(\mathrm{i}\boldsymbol{G}\boldsymbol{R}) \tag{1.1.4}$$

is used, where $\boldsymbol{G}$ are surface reciprocal lattice vectors and $V_{\boldsymbol{G}}$ are the diffractive potentials

$$V_{\boldsymbol{G}}(z) = \frac{1}{\sigma}\int_\sigma V_s(\boldsymbol{r})\exp(-\mathrm{i}\boldsymbol{G}\boldsymbol{R})\,\mathrm{d}\boldsymbol{R} \tag{1.1.5}$$

(σ is the area of the surface unit cell). As a rule, the approximation $V_0(z) = V_0(z)$, $V_{\boldsymbol{G}}(z) = \kappa_{\boldsymbol{G}} V_1(z)$, $\boldsymbol{G} \neq 0$ is used, $\kappa_{\boldsymbol{G}}^* = \kappa_{-\boldsymbol{G}}$ are the corrugation coefficients. Only a few terms of this expansion are taken into account usually. The functions $V_{0,1}(z)$ are modelled either by Morse type potentials

$$V_0(z) = V_{0\mathrm{M}}(z) \equiv D\left(\mathrm{e}^{-2\lambda z} - 2\mathrm{e}^{-\lambda z}\right), \tag{1.1.6}$$

$$V_1(z) = V_{1\mathrm{M}}(z) \equiv D\left(\mathrm{e}^{-2\lambda z} - \mathrm{e}^{-\lambda z}\right), \tag{1.1.7}$$

or by Born–Mayer potential

$$V_0(z) = V_1(z) = V_{\mathrm{BM}}(z) \equiv A\mathrm{e}^{-2\lambda z} \tag{1.1.8}$$

(the factor 2 is introduced into the exponent of the latter expression to match the two models).

For long wave phonons and a short-range pairwise potential, the energy of atom–phonon interaction is mainly determined by the displacement $\boldsymbol{u}(\boldsymbol{R})$ of surface point $\boldsymbol{R}$ posed exactly below the gas atom: $V(\boldsymbol{r},\{\boldsymbol{u}_k\}) \approx V_s(\boldsymbol{r}) - \boldsymbol{u}(\boldsymbol{R})\nabla V_s(\boldsymbol{r})$, or

$$f_l(\boldsymbol{r}) = -\frac{\boldsymbol{e}_l}{\sqrt{2\omega_l MN}} \exp(-\mathrm{i}\boldsymbol{Q}\boldsymbol{R})\nabla V_s(\boldsymbol{r}). \tag{1.1.9}$$

Of course, this does not assume the interaction with a single atom.

In order to obtain this equation more rigorously, the potential of the gas atom–phonon interaction is assumed to be determined by the sum of the pairwise potentials $v(r)$. In this case inserting the expression $1 = \int \mathrm{d}\boldsymbol{R}'\,\delta(\boldsymbol{R}' - \boldsymbol{R}_k)$ into (1.1.3) gives

$$f_l(\boldsymbol{r}) = \frac{\partial}{\partial \boldsymbol{r}} \int \mathrm{d}\boldsymbol{R}' \exp(-\mathrm{i}\boldsymbol{Q}\boldsymbol{R}')v(|\boldsymbol{r} - \boldsymbol{R}'|)\sum_k \delta(\boldsymbol{R}' - \boldsymbol{R}_k) \tag{1.1.10}$$

and, since $\sigma \sum_k \delta(\boldsymbol{R}' - \boldsymbol{R}_k) = \sum_n \exp(\mathrm{i}\boldsymbol{G}_n \boldsymbol{R}')$,

$$f_l(\boldsymbol{r}) = \frac{\partial}{\partial \boldsymbol{r}} \sum_n \int \frac{\mathrm{d}\boldsymbol{R}'}{\sigma} \exp[-\mathrm{i}(\boldsymbol{G}_n - \boldsymbol{Q})\boldsymbol{R}'] v(|\boldsymbol{r} - \boldsymbol{R}'|)$$
$$= \frac{\partial}{\partial \boldsymbol{r}} \sum_n V_{\boldsymbol{G}-\boldsymbol{Q}}(z) \exp[\mathrm{i}(\boldsymbol{G} - \boldsymbol{Q})\boldsymbol{R}], \qquad (1.1.11)$$

where $V_{\boldsymbol{G}-\boldsymbol{Q}}(z)$ is determined by (1.1.5). This function has a sharp maximum at the origin. Assuming $V_{\boldsymbol{Q}}(z) = V_0(z)\exp(-Q^2/Q_c^2)$ yields the wave vector cutoff (Bortolani et al. 1983). For long wave phonons with $Q \ll G$ the approximation $V_{\boldsymbol{G}-\boldsymbol{Q}} \approx V_{\boldsymbol{G}}$ gives (1.1.9) again.

The presence in (1.1.9) of the Bloch factor $\exp(-\mathrm{i}\boldsymbol{Q}\boldsymbol{R})$ describes the translation symmetry of the interaction, and allows one to take into account tangential momentum exchange, closely associated with this symmetry. Owing to this factor the effective frequency of adatom–phonon mode interaction depends on the direction of the wave vector.

When the scattered particle has an internal structure, the potential should take into account the corresponding degrees of freedom (Gorbachev et al. 1991). In this review the consideration is limited to the case of diatomic molecules modelled by plane rotors. In this case the static potential (1.1.4) depends on molecular internal variables as well. Usually this dependence is approximated by several terms of the expansion over the Legendre polynomials of the cosine of the rotational variable θ and in powers of the vibrational variable $\rho - \rho_0$ (θ is the angle between the molecular axis and the surface normal, while ρ is the interatomic distance with its equilibrium value ρ_0):

$$V_s(\boldsymbol{r}, \theta, \rho) = V_0(z) + \sum_{l,n,\boldsymbol{G}} V_{\boldsymbol{G}}^{ln}(z)\, P_l(\cos\theta)(\rho - \rho_0)^n \exp(\mathrm{i}\boldsymbol{G}\boldsymbol{R}). \qquad (1.1.12)$$

Owing to the lack of information on the coefficients $V_{\boldsymbol{G}}^{ln}(z)$, they are assumed to be proportional to $V_{\boldsymbol{G}}(z)$

$$V_{\boldsymbol{G}}^{ln}(z) = a_{\boldsymbol{G}}^{ln} \kappa_{\boldsymbol{G}} V_1(z), \qquad (1.1.13)$$

with anisotropy coefficients $a_{\boldsymbol{G}}^{ln}$ rapidly decreasing with their indices. It is these anisotropy coefficients that are responsible for different resonances and mutual coupling effects, namely the resonances between rotations and vibrations of the molecule, thermal vibrations of the surface, and the motion along the crystal "washboard".

1.2 Path Integral Representation of S-matrix

The S-matrix of gas atom and crystal at finite times reads

$$\langle \boldsymbol{p}_\mathrm{f}\, \{n_{l\mathrm{f}}\} \,|\, \mathrm{S} \,|\, \boldsymbol{p}_\mathrm{i}\, \{n_{l\mathrm{i}}\} \rangle = \langle \mathrm{f} \,|\, \mathrm{e}^{\mathrm{i}H_0 t_\mathrm{f}} \mathrm{e}^{-\mathrm{i}H(t_\mathrm{f} - t_\mathrm{i})} \mathrm{e}^{-\mathrm{i}H_0 t_\mathrm{i}} \,|\, \mathrm{i} \rangle, \qquad (1.2.1)$$

where $\boldsymbol{p}$ is the momentum of the gas atom, $n_{l\mathrm{f,i}}$ are the sets of the occupation numbers of crystal normal modes l, indices i and f denote the initial and final states of the scattering. H_0 is the Hamiltonian of free motion, which does not involve the potential of the gas atom–surface interaction. The corresponding classical Hamiltonians are

$$H_0 = \frac{p^2}{2m_g} + \sum_l \omega_l a_l^* a_l, \tag{1.2.2}$$

where m_g is the mass of the gas atom, and

$$H = H_0 + V_s(r) - \sum_l [f_l^*(r)a_l + f_l(r)a_l^*]. \tag{1.2.3}$$

The S-matrix may be expressed through the conventional path integral in coordinate representation by the Fourier transform in $p_{f,i}$ (Bogdanov et al. 1989):

$$\langle f \,|\, S(t_f, t_i) \,|\, i \rangle = \exp\left(i \frac{p^2}{2m_g} t \Big|_{t=t_i}^{t_f} \right) \int dr_f \, dr_i \, \langle p_f \,|\, r_f \rangle \\ \times \langle r_f \{n_{lf}\} \,|\, \exp[-iH(t_f - t_i)] \,|\, r_i \{n_{li}\} \rangle \langle r_i \,|\, p_i \rangle, \tag{1.2.4}$$

where

$$\langle r_f \{n_{lf}\} \,|\, \exp[-iH(t_f - t_i)] \,|\, r_i \{n_{li}\} \rangle \\ = \int \mathcal{D}r \, \mathcal{D}p \prod_l G_{fi}^l[r(t)] \exp\left[i \int_{r_i}^{r_f} p \, dr - i \int_{t_i}^{t_f} \left(\frac{p^2}{2m_g} + V_s(r) \right) dt \right], \tag{1.2.5}$$

$G_{fi}^m[r(t)]$ stands for the amplitude of the i $\to$ f transition in phonon mode m. Analogous expressions for the scattering amplitude in coordinate representation were written by different authors, to mention (Pechukas and Davis 1972; Newns 1985; Nourtier 1985; Billing 1990) for example.

After integration by parts in the exponent, this expression becomes regular in the limit $t_i \to -\infty$, $t_f \to +\infty$. Including the integration over r_i, r_f into the measure yields

$$\langle f \,|\, S \,|\, i \rangle = (2\pi)^{-3} \int_{p_i}^{p_f} \mathcal{D}r \, \mathcal{D}p \prod_l G_{fi}^l[r(t)] \exp\{iS[r(t), p(t)]\}, \tag{1.2.6}$$

where

$$S[r(t), p(t)] = -\int_{p_i}^{p_f} \left(r - \frac{p}{m_g} t \right) dp - \int_{-\infty}^{+\infty} V_s(r) \, dt \tag{1.2.7}$$

is the classic action of the gas atom, and the path integral is defined now as

$$\int_{p_i}^{p_f} \mathcal{D}r \, \mathcal{D}p = \lim_{N \to \infty} \int \ldots \int dr_1 \frac{dp_{1/2}}{(2\pi)^3} dr_2 \ldots \frac{dp_{N-1/2}}{(2\pi)^3} dr_N. \tag{1.2.8}$$

The trajectories in this path integral meet boundary conditions $p(-\infty) = p_i$, $p(+\infty) = p_f$.

The amplitude $\prod_l G_{fi}^l$ may be evaluated in the Bargmann–Fock representation, then the S-matrix may be written as

$$\langle f \,|\, S(t_f, t_i) \,|\, i \rangle = \int \left(\prod_l \frac{d^2 a_l}{\pi} \frac{d^2 \alpha_l}{\pi} \exp\left(-|a_l|^2 - |\alpha_l|^2 \right) \langle n_{lf} \,\|\, a_l \rangle \langle \alpha_l \,\|\, n_{li} \rangle \right) \\ \times \langle p_f \{a_l\} \,|\, S(t_f, t_i) \,|\, p_i \{\alpha_l\} \rangle, \tag{1.2.9}$$

where $\| a \rangle$ and $\| \alpha \rangle$ are the Bargmann states of phonon oscillators:

$$\| a\rangle = \sum_{n=0}^{\infty} \frac{a^n}{\sqrt{n!}} | n\rangle, \qquad \langle a \| a\rangle = \mathrm{e}^{a^* a}. \tag{1.2.10}$$

In this representation the S-matrix has the form of a path integral over variables $(\boldsymbol{r}, \boldsymbol{p})$ and (a^*, a) (Faddeev and Slavnov 1980):

$$\langle \boldsymbol{p}_\mathrm{f} \{a_l\} \,|\, \mathsf{S}(t_\mathrm{f}, t_\mathrm{i}) \,|\, \boldsymbol{p}_\mathrm{i} \{\alpha_l\}\rangle = \int_{\boldsymbol{p}_\mathrm{i}}^{\boldsymbol{p}_\mathrm{f}} \frac{\mathcal{D}\boldsymbol{r}\, \mathcal{D}\boldsymbol{p}}{(2\pi)^3} \prod_l \mathsf{S}_l[a_l^*, \alpha_l; \boldsymbol{r}(t)] \exp\{\mathrm{i}S[\boldsymbol{r}(t), \boldsymbol{p}(t)]\}. \tag{1.2.11}$$

The S-matrix of lth phonon mode S_l may be represented by the path integral over complex variables (a, a^*):

$$\mathsf{S}_l\left[a_l^*, \alpha_l; \boldsymbol{r}(t)\right] = \int \mathcal{D}a^*\, \mathcal{D}a \exp(\mathrm{i}S_{\mathrm{osc}}), \tag{1.2.12}$$

where

$$S_{\mathrm{osc}}[a(t), a^*(t)] = \frac{1}{2\mathrm{i}}\left[a_l^* a(t_\mathrm{f})\mathrm{e}^{\mathrm{i}\omega_l t_\mathrm{f}} + \alpha_l a^*(t_\mathrm{i})\mathrm{e}^{-\mathrm{i}\omega_l t_\mathrm{i}}\right] + \int_{t_\mathrm{i}}^{t_\mathrm{f}} \left(\frac{\dot{a}^* a - a^* \dot{a}}{2\mathrm{i}} - \omega_l a^* a + f_l a^* + f_l^* a\right) \mathrm{d}t. \tag{1.2.13}$$

Trajectories $a(t)$ and $a^*(t)$ in the path integral meet the boundary conditions $a(t_\mathrm{i}) = \alpha_l \exp(-\mathrm{i}\omega_l t_\mathrm{i})$, $a^*(t_\mathrm{f}) = a_l^* \exp(\mathrm{i}\omega_l t_\mathrm{f})$ respectively. This path integral is well known (Faddeev and Slavnov 1980):

$$\mathsf{S}_l\left[a_l^*, \alpha_l; \boldsymbol{r}(t)\right] = G_{00}^l[\boldsymbol{r}(t)] \exp\left(a_l^* \alpha_l + \mathrm{i}\beta_l a_l^* + \mathrm{i}\beta_l^* \alpha_l\right), \tag{1.2.14}$$

where

$$\beta_l = \int f_l(t) \mathrm{e}^{\mathrm{i}\omega_l t}\, \mathrm{d}t \tag{1.2.15}$$

is the Fourier transform of the force at a trajectory $\boldsymbol{r}(t)$. This expression is regular in the limit $t_\mathrm{i} \to -\infty$, $t_\mathrm{f} \to +\infty$. Substition of (1.2.11), (1.2.13), and energy states in the coherent basis

$$\langle a \| n\rangle = \frac{(a^*)^n}{\sqrt{n!}} \tag{1.2.16}$$

into (1.2.9) and integration over $\mathrm{d}^2 a_l\, \mathrm{d}^2 \alpha_l$ complete the consideration:

$$G_{\mathrm{fi}}^l = \frac{G_{00}^l}{\sqrt{n_{l\mathrm{f}}! n_{l\mathrm{i}}!}} \sum_{r=0}^{\min(n_{l\mathrm{f}}, n_{l\mathrm{i}})} \binom{n_{l\mathrm{f}}}{r} \binom{n_{l\mathrm{i}}}{r} r!\, (\mathrm{i}\beta_l)^{n_{l\mathrm{f}}-r} (\mathrm{i}\beta_l^*)^{n_{l\mathrm{i}}-r}, \tag{1.2.17}$$

with

$$G_{00}^l = \exp\left[-\frac{1}{2} \iint \mathrm{d}s\, \mathrm{d}t\, f_l^*(s) f_l(t) \mathrm{e}^{-\mathrm{i}\omega_l |t-s|}\right] \tag{1.2.18}$$

being the vacuum–vacuum amplitude.

The scattering probability may be obtained by averaging $|\langle \mathrm{f} \,|\, \mathsf{S} \,|\, \mathrm{i}\rangle|^2$ over initial states of the crystal and the summation over final ones:

$$P(\boldsymbol{p}_\mathrm{f}, \boldsymbol{p}_\mathrm{i}) = (2\pi)^{-6} \int_{\boldsymbol{p}_\mathrm{i}}^{\boldsymbol{p}_\mathrm{f}} \mathcal{D}\Gamma\, \mathcal{D}\Gamma'\, F[\Gamma, \Gamma'] \exp\{\mathrm{i}S[\Gamma] - \mathrm{i}S[\Gamma']\}, \tag{1.2.19}$$

where $\int \mathcal{D}\Gamma$ denotes the path integral over classical trajectories of the gas atom $\Gamma(t) = \{\boldsymbol{r}(t), \boldsymbol{p}(t)\}$, and $F[\Gamma, \Gamma']$ is the influence functional (Feynman and Hibbs 1965; Grabert et al. 1988) describing the surface influence on the particle:

$$F[\boldsymbol{r}(t), \boldsymbol{r}'(t)] = \prod_l \sum_{\{n_{lf}, n_{li}\}} G^m_{\mathrm{fi}}\, G^{l\,\prime *}_{\mathrm{fi}} \exp\left(-\frac{n_{li}\omega_l}{k_{\mathrm{B}}T_{\mathrm{s}}}\right)\left[1 - \exp\left(-\frac{\omega_l}{k_{\mathrm{B}}T}\right)\right]$$
$$= \prod_l G^l_{00}\, G^{l\,\prime *}_{00} \exp\left(\beta_l \beta_l'^* - n_l \left|\beta_l - \beta_l'\right|^2\right). \tag{1.2.20}$$

Here k_{B} is the Boltzmann constant, T_{s} is the surface temperature, and $n_l = (\exp(\omega_l/k_{\mathrm{B}}T_{\mathrm{s}}) - 1)^{-1}$ is the average population number of phonon mode l. By virtue of equation $f(-\boldsymbol{q}, j) = f(\boldsymbol{q}, j)^*$ the influence functional F may be written in the form

$$F[\boldsymbol{r}(t), \boldsymbol{r}'(t)]$$
$$= \exp\left\{-\sum_l \int \mathrm{d}t \int^t \mathrm{d}s\, [f_l(t) - f_l'(t)]\,[f_l(s)\gamma_l^*(s-t) - f_l'(s)\gamma_l(s-t)]^*\right\}, \tag{1.2.21}$$

where $\gamma_l(t) = (n_l + 1)\mathrm{e}^{\mathrm{i}\omega_l t} + n_l \mathrm{e}^{-\mathrm{i}\omega_l t}$ is the Green function of oscillator at the temperature T_{s}.

1.3 Stationary Phase Method for Path Integrals

Path integrals in (1.2.19) cannot be evaluated directly by the stationary phase method because classical trajectories minimizing the action are degenerate. This degeneracy is due to the symmetry of the problem: such trajectories may be shifted in time and translated along the surface. Another difficulty consists in essential singularity of $|\langle \mathrm{f} \,|\, \mathrm{S} \,|\, \mathrm{i} \rangle|^2$. It contains δ-functions of energy and quasimomentum conservation.

Both difficulties can be overcome by the method originally introduced in the quantum field theory (Popov 1983) and applied later to the quasiclassical theory of molecule–molecule and gas–surface scattering (Dubrovskiy and Bogdanov 1979a; Bogdanov 1980; Bogdanov et al. 1989).

In case of gas-surface scattering, the stationary phase trajectories in th path integrals (1.2.19) can be fixed both in time and in space by using the following unity decompositions:

$$1 = \int \mathrm{d}\boldsymbol{R}_0\, \mathrm{d}\boldsymbol{R}_0'\, \delta\big(\boldsymbol{R}(\tau') - \boldsymbol{R}_0\big)\delta\big(\boldsymbol{R}'(\tau' - \tau) - \boldsymbol{R}_0'\big), \tag{1.3.1}$$

$$1 = \dot{p}_z(t)\dot{p}_z'(t') \int \mathrm{d}\tau\, \mathrm{d}\tau'\, \delta\big(p_z(\tau')\big)\delta\big(p_z'(\tau' - \tau)\big), \tag{1.3.2}$$

with t, t' being the time moments corresponding to the turning points of the trajectories Γ, Γ'. After the variables transform $\Gamma(t) \to \Gamma(t-\tau')$, $\Gamma'(t) \to \Gamma(t+\tau-\tau')$ the classical trajectory problem for the path integrals (1.2.19) becomes nondegenerate because δ-functions $\delta\big(p_z(0)\big)\delta\big(\boldsymbol{R}(0) - \boldsymbol{R}_0\big)$ in each path integral define the additional boundary conditions for the trajectories at $t = 0$: $p_z(\pm 0) = 0$, $\boldsymbol{R}(\pm 0) = \boldsymbol{R}_0$. It is obvious from (1.2.7), (1.2.21) that

$$F[\Gamma(t-\tau'), \Gamma'(t+\tau-\tau')] = F[\Gamma(t), \Gamma'(t+\tau)], \tag{1.3.3}$$

$$S[\Gamma(t-\tau')] - S[\Gamma'(t+\tau-\tau')] = S[\Gamma(t)] - S[\Gamma'(t)] + \tau \Delta E, \tag{1.3.4}$$

where $\Delta E = (p_\mathrm{f}^2 - p_\mathrm{i}^2)/2m_g$.

Now one encounters a problem of treating $F[\Gamma, \Gamma']$ in the path integrals. If it is considered merely as a pre-exponential factor, the well known classical trajectory approximation is obtained (Brako and Newns 1982; Newns 1985). But this approximation fails when the energy transfer during the collision is comparatively large, besides that, it neglects the tangential momentum exchange which determines, for example, the out of incidence plane scattering. The most direct way to evaluate these path integrals is to vary the logarithm of the influence functional (1.2.21) together with the classical action of the particle (1.2.7), but this leads to very complicated non-local generalized Langevin equations of motion and does not allow one to obtain analytical expressions (Newns 1985; Nourtier 1985).

A possible solution to the problem consists in the following procedure in the spirit of the generalized eikonal method (Dubrovskiy and Bogdanov 1979a). The both path integrals are represented as the compositions of integrals over $\mathcal{D}z\,\mathcal{D}p_z$ in momentum representation and integrals over the tangential variables in a mixed coordinate–momentum representation for $t > 0$ and $t < 0$ (the integrals over $\mathrm{d}p_z(0)\,\mathrm{d}\boldsymbol{R}(0)$ are easily performed due to the δ-functions):

$$\begin{aligned} &\int_{\boldsymbol{p}_\mathrm{i}}^{\boldsymbol{p}_\mathrm{f}} \mathcal{D}\boldsymbol{r}\,\mathcal{D}\boldsymbol{p}\,\delta(p_z(0))\,\delta(\boldsymbol{R}(0) - \boldsymbol{R}_0) \\ &\quad = \frac{1}{2\pi} \int_{p_{\mathrm{i}z}}^{p_z(0)=0} \mathcal{D}z\,\mathcal{D}p_z \int_{p_z(0)=0}^{p_{\mathrm{f}z}} \mathcal{D}z\,\mathcal{D}p_z \\ &\qquad \times \int_{\boldsymbol{P}_\mathrm{i}}^{\boldsymbol{R}(0)=\boldsymbol{R}_0} \mathcal{D}\boldsymbol{R}\,\mathcal{D}\boldsymbol{P} \int_{\boldsymbol{R}(0)=\boldsymbol{R}_0}^{\boldsymbol{P}_\mathrm{f}} \mathcal{D}\boldsymbol{R}\,\mathcal{D}\boldsymbol{P}. \end{aligned} \tag{1.3.5}$$

After that each of the integrals is evaluated within the quasiclassical method separately before and after fixing point $t = 0$. The influence functional $F[\Gamma, \Gamma']$ at each trajectory branch is considered as a pre-exponential factor and is replaced with its value at the classical trajectory, obtained by the variation of $S - S'$. In this case the classical trajectories $\boldsymbol{r}(t), \boldsymbol{r}'(t)$ are real and coincide, but they have jumps in z and $\boldsymbol{P}$ at $t = 0$ (while the trajectories of the generalized eikonal method have jumps only in $\boldsymbol{p}$).

This method allows one to take into account the influence of crystal on adatom and can be a starting point for perturbation theory, that would advance this approximation. Of course, it is valid only provided F varies slower than $\exp(\mathrm{i}S)$, but this question is beyond the present discussion.

If surface corrugation is neglected, the potential in the equations of motion depends only on z, and the tangential motion is free at each branch of the trajectory. The classical trajectory $\boldsymbol{R}(t), z(t)$ is then found from the equations of motion involving the only potential V_0 with boundary conditions $m_g\dot{z}(\pm 0) = 0$, $\boldsymbol{R}(\pm 0) = \boldsymbol{R}_0$; $m_g\dot{\boldsymbol{r}} \to \boldsymbol{p}_{\mathrm{f,i}}$ at $t \to \pm\infty$:

$$m_g\ddot{z} = -V_0'(z),$$
$$R(t) = R_0 + [V_i\theta(-t) + V_f\theta(t)]t, \tag{1.3.6}$$

where $\theta(t)$ is the Heaviside function and $v = (V, v_z) = p/m_g$ is the vector of velocity. Of course, this trajectory has a jump at $t = 0$.

This approximation for the classical trajectory is valid only for the scattering in thermal regime, when the energy is not too large. But it can also be used when for the diffractive scattering due to the following reason. The quasiclassical expression for the path integral in mixed representation over tangential part of $\mathcal{D}\Gamma$ merely reads

$$\exp\left\{-i\left[R_0\Delta P(0) + \left(\int_{-\infty}^{-0} + \int_{+0}^{+\infty}\right)\left(R - \frac{P}{m_g}t\right)\dot{P}\,dt\right]\right\}$$

where $\Delta P(0) = P(+0) - P(-0)$. For the trajectory (1.3.6) $\Delta P(0) = \Delta P \equiv P_f - P_i$ and $\dot{P} = 0$. If surface corrugation is taken into account, the determinant remains close to unit. Since the classical action corresponding to the potential V_0 is stationary at the classical trajectory (1.3.6), the action at the perturbed trajectory differs only by the term $\int dt\,[V_s(r) - V_0(z)]$ that can also be evaluated along the trajectory (1.3.6).

1.4 Path Integral over Variables of Normal Motion

For the path integrals over the normal motion variables, the standard quasiclassical approximation cannot be used directly since the quasiclassical amplitude in coordinate representation has a singularity at the turning point (in our case at $t = 0$) (Landau and Lifshitz 1965) because there are, in general, two classical trajectories meeting the boundary condition for coordinates $z(t_1) = z_1$, $z(t_2) = z_2$. As one of the points $z_{1,2}$ tends to the turning point, these trajectories tend to each other and the usual quasiclassical method looses its validity.

The quasiclassical amplitude in the momentum representation does not suffer from this feature, because boundary conditions $p_z(t_1) = p_{1z}$, $p_z(t_2) = p_{2z}$ determine the unique classical trajectory for the typical scattering potentials. Such amplitude cannot be obtained as a Fourier transform of the quasiclassical propagator in coordinate representation. So it is necessary to modify the stationary phase method for the evaluation of the path integral in momentum representation.

In the quasiclassical approximation one has

$$\int_{p_1}^{p_2} \mathcal{D}q\,\mathcal{D}p \exp\{iS[q(t), p(t)]\} = \Delta \exp(iS_{cl}), \tag{1.4.1}$$

whith S_{cl} being the action at the classical trajectory $\{q_{cl}(t), p_{cl}(t)\}$, and

$$\Delta = \int_{p(t_1)=0}^{p(t_2)=0} \mathcal{D}q\,\mathcal{D}p \exp\left\{-i\int_{t_1}^{t_2}\left[q\dot{p} + \frac{p^2}{2m_g} + \frac{v(t)q^2}{2}\right]dt\right\}, \tag{1.4.2}$$

where $v(t) \equiv \partial^2 V(q_{cl}(t))/\partial q^2$. Integration over the coordinates using finite approximation of the path integral yields

$$\Delta = \lim_{N\to\infty} \frac{2\pi}{\sqrt{2\pi i \tau v(t_2)}} \int \prod_{k=1}^{N-1} \frac{dp_k}{\sqrt{2\pi i \tau v_k}} \exp\left[i\sum_{k=1}^{N}\left(\frac{\dot{p}_k^2}{2v_k} - \frac{p_k^2}{2m_g}\right)\right], \tag{1.4.3}$$

where $p_k = p(t^k)$, $v_k = v(t^k)$, $\tau = (t_2 - t_1)/N$ and $t^k = t_1 + k\tau$. This transformation is analogous to the transition from the Hamiltonian form of the path integral to the Lagrangian one.

For the evaluation of this integral the method used in the Appendix 1 to the book by R. Rajaraman (Rajaraman 1982) has been modified. It is convenient to append the integration over the final momentum:

$$\Delta = \int \mathrm{d}p_N\, \delta(p_N)\Delta = \int \mathrm{d}\alpha \int \mathcal{D}_v p \exp\left\{\mathrm{i}\left[\alpha p(t_2) + \int_{t_1}^{t_2}\left(\frac{\dot{p}^2}{2v} - \frac{p^2}{2m_g}\right)\mathrm{d}t\right]\right\}, \tag{1.4.4}$$

where

$$\int \mathcal{D}_v p = \lim_{N\to\infty}\int \frac{\mathrm{d}p_1}{\sqrt{2\pi\mathrm{i}\tau v_1}} \cdots \frac{\mathrm{d}p_N}{\sqrt{2\pi\mathrm{i}\tau v_N}}.$$

In order to get rid of $-p^2/2m_g$ in the exponent the functional change of variable may be used:

$$\eta(t) = p(t) - \int_{t_1}^{t_2} \frac{\dot{f}(\tau)}{f(\tau)} p(\tau)\,\mathrm{d}\tau. \tag{1.4.5}$$

If function f meets the equation

$$\frac{\mathrm{d}}{\mathrm{d}t}\left(\frac{\dot{f}}{v}\right) + \frac{f}{m} = 0 \tag{1.4.6}$$

with boundary condition $\dot{f}(t_2) = 0$ (such f is not unique), then

$$\alpha p(t_2) + \int_{t_1}^{t_2}\left(\frac{\dot{p}^2}{2v} - \frac{p^2}{2m_g}\right)\mathrm{d}t = \int_{t_1}^{t_2}\left[\frac{\dot{\eta}^2}{2v} + \alpha\frac{f(t_2)}{f(t)}\dot{\eta}\right]\mathrm{d}t. \tag{1.4.7}$$

The Jacobian of this variable change is $J = [f(t_2)/f(t_1)]^{1/2}$ (Gel'fand and Yaglom 1960). Thus obtained path integral may be evaluated with the help of finite approximation $\eta_k = \eta(t^k)$: the change of variables $b_1 = \eta_1/\sqrt{v_1}$, $b_k = (\eta_k - \eta_{k-1})/\sqrt{v_k}$, $2 \le k \le N$, with the Jacobian

$$\frac{\partial(b_1, \ldots, b_N)}{\partial(\eta_1, \ldots, \eta_N)} = \prod_{k=1}^{N} \frac{1}{\sqrt{v_k}} \tag{1.4.8}$$

yields

$$\Delta = \left[\frac{\mathrm{i}}{2\pi}F(t_1, t_2)\right]^{-1/2}, \tag{1.4.9}$$

where

$$F(t_1, t_2) = f(t_1)f(t_2)\int_{t_1}^{t_2} \frac{v(\tau)}{f^2(\tau)}\,\mathrm{d}\tau. \tag{1.4.10}$$

Function $F(t_1, t_2)$ does not depend on the boundary conditions for f and may be related to the classical action $S(p_1, p_2)$. It follows from the definition of $F(t_1, t_2)$ that it meets the equation

$$m_g \frac{\partial^2}{\partial t^2}\left(\frac{F}{v}\right) + F = 0 \tag{1.4.11}$$

with the boundary conditions

$$F(t_1,t_1)=0, \qquad \frac{\partial F(t_1,t_1)}{\partial t_2}=v(t_1). \tag{1.4.12}$$

It is easy to see from these relations that

$$F(t_1,t_2)=-\frac{\partial p_2}{\partial q_{\rm cl}(t_1)}=-\left[\frac{\partial q_{\rm cl}(t_2)}{\partial p_1}\right]^{-1}=-\left[\frac{\partial^2 S(p_1,p_2)}{\partial p_1\partial p_2}\right]^{-1}, \tag{1.4.13}$$

hence

$$\Delta=\left[2\pi\mathrm{i}\frac{\partial^2 S(p_1,p_2)}{\partial p_1\partial p_2}\right]^{1/2}. \tag{1.4.14}$$

This expression formally coincides with the Fourier transform of the propagator in coordinate representation. Actually it is equivalent to its analytical continuation in the case when $p_1=0$ or $p_2=0$.

Finally,

$$\int_{t<0}\mathcal{D}z\,\mathcal{D}p_z\exp(\mathrm{i}S_z)=\left[2\pi\mathrm{i}\frac{\partial z(-0)}{\partial p_{\mathrm{i}z}}\right]^{1/2}\exp[\mathrm{i}S_z(t<0)], \tag{1.4.15}$$

$$\int_{t>0}\mathcal{D}z\,\mathcal{D}p_z\exp(\mathrm{i}S_z)=\left[-2\pi\mathrm{i}\frac{\partial z(+0)}{\partial p_{\mathrm{f}z}}\right]^{1/2}\exp[\mathrm{i}S_z(t>0)], \tag{1.4.16}$$

where S_z is the part of the classical action S involving the normal motion variables. The classical trajectory $z(t)$ in (1.4.15) and (1.4.16) has a jump at $t=0$ and consists of two branches corresponding to different momenta $p_{\mathrm{f,i}z}$ when $t=\pm\infty$.

The factor $\dot p_z(0)$ in the path integrals must be evaluated along the continuous trajectories, but we shall take its value as an average over two branches of the trajectory $\langle\dot p_z(0)\rangle$. It is convenient to take for example $\langle\dot p_z(0)\rangle=[\dot p_z(-0)\dot p_z(+0)]^{1/2}$. Hence, by virtue of

$$\begin{aligned}\dot p_z(-0)\frac{\partial z(-0)}{\partial p_{\mathrm{i}z}}&=-V_{\rm s}'(z(-0))\frac{\partial z(-0)}{\partial p_{\mathrm{i}z}}=-\frac{\mathrm{d}}{\mathrm{d}p_{\mathrm{i}z}}V_{\rm s}(z(-0))\\&=-\frac{\mathrm{d}}{\mathrm{d}p_{\mathrm{i}z}}\left(\frac{p_{\mathrm{i}z}^2}{2m_g}\right)=\frac{|p_{\mathrm{i}z}|}{m_g}\end{aligned} \tag{1.4.17}$$

and analogously

$$-\dot p_z(+0)\frac{\partial z(+0)}{\partial p_{\mathrm{f}z}}=\frac{p_{\mathrm{f}z}}{m_g}, \tag{1.4.18}$$

the quasiclassical approximation for the path integrals reads

$$\int_{p_{\mathrm{i}z}}^{p_{\mathrm{f}z}}\mathcal{D}z\,\mathcal{D}p_z\,\delta(p_z(0))\dot p_z(0)\exp(\mathrm{i}S_z)=\frac{\sqrt{|p_{\mathrm{i}z}|p_{\mathrm{f}z}}}{m_g}\exp\left\{\mathrm{i}S_z[z(t)]+\mathrm{i}\frac{\pi}{2}\right\}. \tag{1.4.19}$$

The additional phase $\pi/2$ in the exponent is due to the focal point $t=0$ (Landau and Lifshitz 1965).

Finally, dividing the probability P by the observation time $\int d\tau'$, the flux of incident particles $(2\pi)^{-3}|p_{\mathrm{i}z}|/m_g$, and the surface area Σ one obtains a semiclassical representation of scattering kernel R of the Van Hove type

$$R(\boldsymbol{p}_{\mathrm{f}},\boldsymbol{p}_{\mathrm{i}}) = \frac{p_{\mathrm{f}z}}{m_g}\int\frac{\mathrm{d}\tau}{2\pi}\exp(\mathrm{i}\tau\Delta E)\int\frac{\mathrm{d}\boldsymbol{R}_0\,\mathrm{d}\boldsymbol{R}_0'}{4\pi^2\Sigma}\exp[-\mathrm{i}(\boldsymbol{R}_0-\boldsymbol{R}_0')\Delta\boldsymbol{P}(0)]$$
$$\times F[\boldsymbol{r}(t),\boldsymbol{r}'(t+\tau)]\exp\left\{\mathrm{i}S\left[\boldsymbol{r}(t),m_g\dot{\boldsymbol{r}}(t)\right]-\mathrm{i}S\left[\boldsymbol{r}'(t),m_g\dot{\boldsymbol{r}}'(t)\right]\right\}, \tag{1.4.20}$$

where $\boldsymbol{r}(t)$ is the classical trajectory described above. This expression is very convenient as a starting point for various further approximations.

1.5 Internal Degrees of Freedom of Scattered Particles

Let us discuss now the opportunities to take into account the processes involving internal degrees of freedom of scattered particles. In the simplest case of a rotating harmonic oscillator which is considered here, the Hamiltonian of free motion should be replaced with

$$H_0 = \frac{p^2}{2m_g}+\frac{J^2}{2I}+\sum_m \omega_m a_m^* a_m + \omega_{\mathrm{vib}} a^* a, \tag{1.5.1}$$

where J is the angular momentum, I is the moment of inertia, ω_{vib} is the frequency of the molecule vibration, and a is defined via the generalized coordinate $q_0 = \rho - \rho_0$ and momentum $p_0 = m_0\dot{\rho}$

$$a = \frac{m_0\omega_{\mathrm{vib}} q_0 + \mathrm{i}p_0}{\sqrt{2m_0\omega_{\mathrm{vib}}}} \tag{1.5.2}$$

with m_0 being the reduced mass. The interaction is given by (1.1.12) with $\rho - \rho_0$ being replaced with $(2m_0\omega_{\mathrm{vib}})^{-1/2}(a+a^*)$.

The account of transitions between the vibrational and rotational states of molecule leads to the appearance of two more pre-exponential factors $G_{\mathrm{fi}}^{\mathrm{vib}}[\boldsymbol{r}(t)]$ and $\mathcal{G}_{\mathrm{fi}}^{\mathrm{rot}}[\boldsymbol{r}(t)]$ in the representation (1.2.6).

The first amplitude may be evaluated exactly in the same way as that of phonons. It is given by (1.2.17) with β_{vib} calculated along the classical trajectory.

As for the rotational amplitude $\mathcal{G}_{\mathrm{fi}}^{\mathrm{rot}}$, it may be calculated in the framework of the formalism of path integration in action–angle variables, developed in molecule collision theory (Bogdanov et al. 1989). For plane rotor the action–angle variables coincide with the angular momentum J and the orientation angle θ, and the rotational amplitude is

$$\mathcal{G}_{\mathrm{fi}}^{\mathrm{rot}}[\boldsymbol{r}(t)] = \int_{J_{\mathrm{i}}}^{J_{\mathrm{f}}} \mathcal{D}\theta\,\mathcal{D}J\,\exp\{\mathrm{i}S_{\mathrm{rot}}[J(t),\theta(t),\boldsymbol{r}(t)]\}, \tag{1.5.3}$$

where the classical action is given by

$$S_{\mathrm{rot}} = -\int_{J_{\mathrm{i}}}^{J_{\mathrm{f}}}(\theta - \omega_{\mathrm{rot}}t)\,\mathrm{d}J - \int_{-\infty}^{+\infty} V_{\mathrm{rot}}(\boldsymbol{r},\theta)\,\mathrm{d}t. \tag{1.5.4}$$

with $V_{\mathrm{rot}}(\boldsymbol{r},\theta)$ being the part of the static potential (1.1.12) involving the rotational degrees of freedom, and $\omega_{\mathrm{rot}}(t) = \dot{\theta}$.

The path integral in (1.5.3) may also be evaluated with the help of the generalized eikonal method. Let us consider the case of a homonuclear molecule. Using the decomposition of a unity

$$1 = \int_{-\infty}^{+\infty} d\theta_0\, \delta(\theta(0) - \theta_0) = \sum_{n=-\infty}^{+\infty} \int_0^{\pi} d\theta_0\, \delta(\theta(0) - \theta_0 - \pi n) \tag{1.5.5}$$

(the limits of integration are determined by the symmetry of the molecule; the summation over n corresponds to the interference of the amplitudes of transitions to indistiguishable final states), and treating

$$\exp\left[-i \int_{-\infty}^{+\infty} V_{\text{rot}}(\boldsymbol{r}, \theta)\, dt\right]$$

in (1.5.3) as a pre-exponential factor, after the change of variables $\theta \to \theta - \theta_0 + \pi n$ one obtains

$$\mathcal{G}_{\text{fi}}^{\text{rot}} = :\delta\left(\frac{\Delta J}{2}\right): \int_0^{\pi} d\theta_0\, \exp\left[i\theta_0 \Delta J - i \int_{-\infty}^{+\infty} V_{\text{rot}}(\boldsymbol{r}, \theta_0 + \omega_{\text{rot}}(t)t)\, dt\right], \tag{1.5.6}$$

where

$$:\delta(x) := \sum_{n=-\infty}^{+\infty} \delta(x - n) \tag{1.5.7}$$

indicates the vibrational spectrum. The classical action is calculated along the same classical trajectory that has been mentioned above. The rotational frequency is constant ($\omega_{\text{rot}}(t) = \omega_{\text{rot}\,\text{f,i}}$) along the both trajectory branches.

The final expression for the scattering probability is then factorized into the product of the dynamic structural factor and rotational and vibrational scattering profiles. If the only term of the expansion (1.1.12) with $P_2(\cos\theta)$ is taken into account, the rotational scattering profile reduces to the integral representation of the Bessel function:

$$\left|\mathcal{G}_{\text{fi}}^{\text{rot}}\right|^2 = J_{\Delta J/2}^2(\beta_{\text{rot}}[\boldsymbol{r}(t)]), \tag{1.5.8}$$

where

$$\beta_{\text{rot}} = a_0^{10} \int V_1(z(t)) \cos 2\omega_{\text{rot}}(t)t\, dt, \tag{1.5.9}$$

(an analogous integral with sinus is neglected here). The integral in β_{rot} may be evaluated just as for phonon degrees of freedom (see Sect. 3.1).

In the case of a heteronuclear molecule $\Delta J/2$ and $2\omega_{\text{rot}}$ in (1.5.6)–(1.5.9) should be replaced with ΔJ and ω_{rot} respectively.

Another approach to the definition of path integrals on manifolds has been discussed by V. Popov (Popov 1983).

The effect of anharmonicity of vibrations may be taken into consideration either with the help of the standard perturbation theory (Feynman and Hibbs 1965) or in the frame of quasiclassical approach.

It is also interesting to obtain the well known Bessel approximation for the vibrational profile. In the limit of large enough vibrational quantum numbers and small vibrational energy transfer

$$r \ll n, \qquad r = |n_{\text{f}} - n_{\text{i}}|, \qquad n = \min(n_{\text{f}}, n_{\text{i}}) \tag{1.5.10}$$

the WKB method (see, for example, (Nayfeh 1981)) gives for the Laguerre polynomial $L_n^{(r)}(x)$, which the exact vibrational amplitude may be expressed through, the uniform over $|\beta_{\text{vib}}|^2$ asymptotic formula: up to an inessential phase factor

$$G_{\rm fi}^{\rm vib} = |\beta_{\rm vib}|^r \exp\left(-\frac{|\beta_{\rm vib}|^2}{2}\right)\sqrt{\frac{n!}{(n+r)!}}L_n^{(r)}\left(|\beta_{\rm vib}|^2\right) \sim J_r(2\sqrt{n}|\beta_{\rm vib}|). \quad (1.5.11)$$

For the sake of brevity, only the simplest case of plane rotating oscillator has been considered here. Nevertheless, more involved situations like polyatomic molecules modelled with non-rigid asymmetric tops with the account of intramolecular vibrational-vibrational and vibrational-rotational couplings, may be studied in the framework of the same formalism (Dubrovskiy et al. 1983; Bogdanov et al. 1985b). Some of the cases where cross effects determine the physical picture of scattering are discussed below (Sect. 2.4).

2 Dynamic Structural Factors of Phonon Scattering

2.1 Dynamic Structural Factor in Thermal Regime

When the energy of incident particle is not too large, the potential surfaces may be assumed to be plane. It means that $V_s(\boldsymbol{r}) = V_0(z)$. In this approximation by virtue of (1.3.6) one obtains

$$R(\boldsymbol{p}_{\rm f}, \boldsymbol{p}_{\rm i}) = \frac{p_{{\rm f}z}}{m_g} D(\Delta \boldsymbol{P}, \Delta E). \quad (2.1.1)$$

The dynamic structural factor D reads

$$D(\Delta \boldsymbol{P}, \Delta E) = \int \frac{{\rm d}\tau}{2\pi} \exp({\rm i}\tau\Delta E) \int \frac{{\rm d}\boldsymbol{R}_0}{4\pi^2} \exp({\rm i}\boldsymbol{R}_0\Delta \boldsymbol{P}) F[\boldsymbol{r}(t), \boldsymbol{r}(t+\tau)]$$
$$= \exp[-2W(\boldsymbol{\zeta})] \int \frac{{\rm d}\boldsymbol{\eta}}{(2\pi)^3} \exp\left[{\rm i}\boldsymbol{\zeta}\boldsymbol{\eta} + K(\boldsymbol{\zeta}, \boldsymbol{\eta})\right], \quad (2.1.2)$$

where $\boldsymbol{\eta}$ denotes a "vector" $(\boldsymbol{R}_0, \tau)$, and $\boldsymbol{\zeta} = (\Delta \boldsymbol{P}, \Delta E)$ is a "vector" of tangential momentum and energy gains (these quantities transform as vectors only for rotations in the plane of the surface). The Debye–Waller exponent $W(\boldsymbol{\zeta})$ describes the attenuation of the specular peak.

The expressions for $W(\boldsymbol{\zeta})$ and $K(\boldsymbol{\zeta}, \boldsymbol{\eta})$ follow from (1.2.21):

$$W(\boldsymbol{\zeta}) = \sum_l |\beta_l(\boldsymbol{\zeta})|^2 (n_l + 1/2), \quad (2.1.3)$$

$$K(\boldsymbol{\zeta}, \boldsymbol{\eta}) = \sum_l |\beta_l(\boldsymbol{\zeta})|^2 [(n_l+1)\exp({\rm i}\boldsymbol{\eta}\boldsymbol{\xi}_l) + n_l \exp(-{\rm i}\boldsymbol{\eta}\boldsymbol{\xi}_l)], \quad (2.1.4)$$

where $\boldsymbol{\xi}_l$ denotes a "vector" $(\boldsymbol{Q}, \omega_l)$ of tangential wave vector and frequency.

It is useful to define a functional

$$B^{\pm}\left[f(z), \Omega\right] = \int_{\substack{t>0 \\ t<0}} {\rm d}t\, f\left(z(t)\right) \exp({\rm i}\Omega t) \quad (2.1.5)$$

trough which $\beta_l = \beta_l^+ + \beta_l^-$ is expressed: $\beta_l^{\pm} = (2\omega_l M N)^{-1/2} e_{lz} B^{\pm}\left[-V_0'(z), \omega_l^{\pm}\right]$. The effective frequency $\omega_l^{\pm} = \omega_l - \boldsymbol{Q}\boldsymbol{V}_{\rm f,i}$ appears due to the Doppler effect. This frequency shift causes the tangential momentum exchange between phonons and gas particle.

For both Morse and Born-Mayer potentials $-V_0'(z) = 2\lambda V_1(z)$. Hence for the Born-Mayer potential

$$\mathrm{Re}\, B^{\pm}\left[V_{\mathrm{BM}}(z), \omega_l^{\pm}\right] = (2\lambda)^{-1} |p_{\mathrm{f,i}z}| \frac{\Lambda_l^{\pm}/2}{\sinh\left(\Lambda_l^{\pm}/2\right)}, \tag{2.1.6}$$

where $\Lambda_l^{\pm} = p_l^{\pm}/|p_{\mathrm{f,i}z}|$ are adiabatic parameters involving characteristic momenta of mode m: $p_l^{\pm} = m_g \pi \omega_l^{\pm}/\lambda$. The imaginary part of β_l cannot be evaluated analytically, but it is easy to obtain its Padé approximant (Blinov and Kulginov 1991):

$$\mathrm{Im}\, B^{\pm}\left[V_{\mathrm{BM}}(z), \omega_l^{\pm}\right] = p_{\mathrm{f,i}z} \frac{2\Lambda_l^{\pm}}{\frac{2}{\pi}(\Lambda_l^{\pm})^2 - \frac{5}{2}\Lambda_l^{\pm} + \frac{9\pi}{2}}. \tag{2.1.7}$$

At present it is not clear enough wheather $\mathrm{Im}\, B^{\pm}$ really contributs to the scattering probability, because when the trajectory is continuous, the imaginary part of β_l often can be neglected. To clarify this question, it is necessary to study the generalized Langevin equations for the classical trajectory.

For the Morse potential

$$\begin{aligned} 2\lambda \mathrm{Re}\, B^{\pm}\left[V_{\mathrm{M1}}(z), \omega_l^{\pm}\right] &= \mathrm{Re}\, B^{\pm}\left[-V_{\mathrm{M0}}'(z), \omega_l^{\pm}\right] \\ &= |p_{\mathrm{f,i}z}| \frac{\Lambda_l^{\pm}}{\sinh \Lambda_l^{\pm}} \cosh\left[\Lambda_l^{\pm}\left(1 - \frac{\alpha^{\pm}}{\pi}\right)\right], \end{aligned} \tag{2.1.8}$$

where $\alpha^{\pm} = \arctan(|p_{\mathrm{f,i}z}|/p_D)$ and $p_D = (2m_g D)^{1/2}$. As the well depth $D \to 0$, expression (2.1.8) tends to (2.1.6). The evaluation of $\mathrm{Im}\, B$ in this case is analogous to the Born-Mayer potentail case.

Two important limits are worth noting here. Firstly, as $\lambda \to \infty$ (the hard wall approximation), the dependence of β_l on $\boldsymbol{P}_{\mathrm{f,i}}$ disappears ($\beta_l \to e_{lz}\Delta p_z/\sqrt{2\omega_l MN}$ for both potentials), and the average parallel momentum transfer is zero. Secondly, when β_l tends to zero (elastic scattering), $D(\zeta)$ tends to $\delta(\Delta \boldsymbol{P})\delta(\Delta E)$.

It is interesting to compare (2.1.8) with commonly (and successfully) used expression for $B^{\pm}$ for the hard wall potential with Beeby correction (Beeby 1971; Brako and Newns 1982; Gibson and Sibener 1988; Manson 1991; Celli et al. 1991):

$$B^{\pm}[-V_0'(z)] = \left(p_{\mathrm{f,i}z}^2 + 2m_g U\right)^{1/2}, \tag{2.1.9}$$

where U stands for the "attractive well depth". This expression is unsatisfactory because of two reasons. On one hand, it is based on the assumption that collision is fast, while for most thermal atomic beams the collision time is of the order of surface oscillation period. On the other hand, the size of the interatomic potential well is comparable to that of the atom-surface averaged potential, hence one cannot say that the gas atom having been accelerated in the well interacts with phonons.

In the limit $|p_{\mathrm{f,i}z}| \to \infty$ from (2.1.8) one gets $B^{\pm} \approx |p_{\mathrm{f,i}z}|$ just as from (2.1.9). But when $|p_{\mathrm{f,i}z}| \to 0$, (2.1.8) gives a Landau-Zener type formula (Zener 1931; Landau and Teller 1936)

$$B^{\pm}[-V_0'(z), \omega_m] \approx p_l^{\pm} \exp\left(-\frac{m_g \omega_l^{\pm}}{\lambda p_D}\right) = \left(2m_g U_l^{\pm}\right)^{1/2}, \tag{2.1.10}$$

where

$$U_l^{\pm} = \frac{(p_l^{\pm})^2}{2m_g} \exp\left(-\frac{2p_l^{\pm}}{\pi p_D}\right), \tag{2.1.11}$$

while (2.1.9) gives $B^{\pm}[-V_0'(z)] \approx (2m_g U)^{1/2}$. Numerical calculations show that the difference between formulae (2.1.8) and (2.1.9) is only few percent for any $p_{\mathrm{f,i}z}$ provided $U = U_m^{\pm}$.

By a happy (or unhappy) accident, the values assigned to U by fitting experimental data are often of the same order as $U_l^{\pm}$ and D. For example, for 18 meV He beam scattering from Ar mono-layer adsorbed at Ag(111) (Gibson and Sibener 1988) $D \approx 8\,\mathrm{meV}$ (Chung et al. 1985, 1986), U is about 12 meV, and $U_l^{\pm} \approx 12\,\mathrm{meV}$ (for modes in the center of the surface Brillouin zone). This may be a reason of wide applicability of the expression (2.1.9).

2.2 Inelastic Scattering in Diffraction Regime

When the potential $V_s(\boldsymbol{r})$ is periodic, the integration over $\boldsymbol{R}_0, \boldsymbol{R}_0'$ in (1.4.20) can be done partially in the explicit form. The symmetry of $V_s(\boldsymbol{r})$ and $F[\boldsymbol{r}, \boldsymbol{r}']$ with respect to translating $\boldsymbol{r}$ by surface unit cell vectors $\boldsymbol{a}_1, \boldsymbol{a}_2$ allows one to replace the integral $\int \mathrm{d}\boldsymbol{R}_0\, \mathrm{d}\boldsymbol{R}_0'$ by sums of the integrals over surface unit cells σ_k. The change of variables $\boldsymbol{R}_0 \to \boldsymbol{R}_0 + \boldsymbol{a}_k$ with $\boldsymbol{a}_k = k_1\boldsymbol{a}_1 + k_2\boldsymbol{a}_2$ in each of the integrals and the summation over one of the translations yields

$$\begin{aligned} R(\boldsymbol{p}_{\mathrm{f}}, \boldsymbol{p}_{\mathrm{i}}) &= \frac{p_{\mathrm{f}z}}{m_g} \frac{\sigma}{4\pi^2} \sum_k \exp(\mathrm{i}\boldsymbol{a}_k \Delta\boldsymbol{P}) \mathcal{H}_k(\Delta\boldsymbol{P}, \Delta E) \\ &= \frac{p_{\mathrm{f}z}}{m_g} \sum_{\boldsymbol{G}} H_{\boldsymbol{G}}(\Delta\boldsymbol{P} - \boldsymbol{G}, \Delta E), \end{aligned} \tag{2.2.1}$$

where

$$\begin{aligned} \mathcal{H}_k(\Delta\boldsymbol{P}, \Delta E) = \int \frac{\mathrm{d}\tau}{2\pi} \exp(\mathrm{i}\tau \Delta E) \int_\sigma \frac{\mathrm{d}\boldsymbol{R}_0\, \mathrm{d}\boldsymbol{R}_0'}{\sigma^2} \exp[-\mathrm{i}(\boldsymbol{R}_0 - \boldsymbol{R}_0')\Delta\boldsymbol{P}] \\ \times F[\boldsymbol{r}(t), \boldsymbol{r}'(t+\tau) + \boldsymbol{a}_k] \exp\{\mathrm{i}S[\boldsymbol{r}] - \mathrm{i}S[\boldsymbol{r}']\} \end{aligned} \tag{2.2.2}$$

and

$$H_{\boldsymbol{G}}(\Delta\boldsymbol{P}, \Delta E) = \frac{\sigma}{4\pi^2} \sum_k \exp(\mathrm{i}\boldsymbol{a}_k \Delta\boldsymbol{P}) \mathcal{H}_k(\Delta\boldsymbol{P} + \boldsymbol{G}, \Delta E). \tag{2.2.3}$$

The quantity $H_{\boldsymbol{G}}(\Delta\boldsymbol{P}, \Delta E)$ is defined in the first Brillouin zone and can be referred to as a general structure and form factor of the nth diffractive peak. In case of elastic scattering the influence functional $F = 1$. The integration over τ and the summation over $\boldsymbol{a}_k$ provide then the energy and quasimomentum conservation:

$$H_{\boldsymbol{G}}(\Delta\boldsymbol{P}, \Delta E) = |T_{\Delta\boldsymbol{P}+\boldsymbol{G}}|^2 \delta(\Delta\boldsymbol{P}) \delta(\Delta E), \tag{2.2.4}$$

where $T_{\Delta\boldsymbol{P}}$ are the diffraction amplitudes

$$T_{\Delta\boldsymbol{P}} = \int_\sigma \frac{\mathrm{d}\boldsymbol{R}_0}{\sigma} \exp\{-\mathrm{i}\boldsymbol{R}_0 \Delta\boldsymbol{P} + \mathrm{i}S[\boldsymbol{r}(t)]\}, \qquad \boldsymbol{R}(0) = \boldsymbol{R}_0, \tag{2.2.5}$$

for which the well known Bessel approximation has been developed (Billing 1975; Bogdanov 1980).

The influence functional in (2.2.2) may be written as

$$F[\boldsymbol{r}(t), \boldsymbol{r}'(t+\tau)+\boldsymbol{a}_k] = \exp[-W(\zeta;\boldsymbol{R}_0) - W^*(\zeta;\boldsymbol{R}_0') + K(\zeta,\boldsymbol{\eta}_k;\boldsymbol{R}_0,\boldsymbol{R}_0')], \quad (2.2.6)$$

where $\boldsymbol{\eta}_k = (\boldsymbol{a}_k, \tau)$,

$$W(\zeta;\boldsymbol{R}_0) = \sum_l \{n_m|\beta_l|^2 - \ln G_{00}^l[\boldsymbol{r}(t)]\}, \quad (2.2.7)$$

$$K(\zeta,\eta;\boldsymbol{R}_0,\boldsymbol{R}_0') = \sum_l [(n_l+1)\beta_l\beta_l'^* \exp(\mathrm{i}\eta\xi_l) + n_l\beta_l^*\beta_l' \exp(-\mathrm{i}\eta\xi_l)], \quad (2.2.8)$$

and

$$\beta_l = \beta_l(\boldsymbol{R}_0) = \beta_l[\boldsymbol{r}(t)], \qquad \beta_l' = \beta_l(\boldsymbol{R}_0') = \beta_l[\boldsymbol{r}'(t)]. \quad (2.2.9)$$

In some cases the influence functional F may be taken out of the integrals over the unit cell in (2.2.2). It is possible either when $\beta_l(\boldsymbol{R}_0)$ and $G_{00}^l(\boldsymbol{R}_0)$ depend on $\boldsymbol{R}_0$ weakly compared to $\boldsymbol{G}\boldsymbol{R}_0$ (that is, when the crystal corrugation is much smaller than the phonon relief), or when $S(\boldsymbol{R}_0)$ oscillates strongly enough for the evaluation of the integrals by the stationary phase method (the classical limit). In the latter case F is treated as a pre-exponential factor and may be replaced with its value at the stationary phase point. Then the general structural factor $H_{\boldsymbol{G}}$ can be factorized into the product of the form factor (diffraction intensities) and the dynamic structural factor:

$$R(\boldsymbol{p}_\mathrm{f},\boldsymbol{p}_\mathrm{i}) = \frac{p_{\mathrm{f}z}}{m_g}\sum_{\boldsymbol{G}} |T_{\Delta\boldsymbol{P}}|^2 D_{\boldsymbol{G}}(\Delta\boldsymbol{P}-\boldsymbol{G},\Delta E), \quad (2.2.10)$$

where

$$D_{\boldsymbol{G}}(\Delta\boldsymbol{P},\Delta E) = \exp[-2W(\zeta)]\frac{\sigma}{4\pi^2}\sum_k \int \frac{\mathrm{d}\tau}{2\pi} \exp[\mathrm{i}\zeta\boldsymbol{\eta}_k + K(\zeta,\boldsymbol{\eta}_k)]. \quad (2.2.11)$$

When the dependence of the exponent on k is weak enough, the summation over k may be replaced by integration. Then the scattering kernel reads

$$R(\boldsymbol{p}_\mathrm{f},\boldsymbol{p}_\mathrm{i}) = \frac{p_{\mathrm{f}z}}{m_g}\sum_{\boldsymbol{G}} |T_{\boldsymbol{G}}|^2 D(\Delta\boldsymbol{P}-\boldsymbol{G},\Delta E), \quad (2.2.12)$$

where $D(\boldsymbol{\zeta})$ is given by (2.1.2). Finally, when D has no sharp peak in $\Delta\boldsymbol{P}$ (compared to the surface first Brillouin zone size), the scattering kernel (2.2.12) reduces to

$$R(\boldsymbol{p}_\mathrm{f},\boldsymbol{p}_\mathrm{i}) = \frac{p_{\mathrm{f}z}}{m_g}|T_{\Delta\boldsymbol{P}}|^2 D(\Delta\boldsymbol{P},\Delta E). \quad (2.2.13)$$

2.3 Single Phonon Approximation

The single phonon approximation is obtained by expanding the exponential expression for the influence functional F into series and keeping two terms only. The fist term describes the elastic scattering while the second one corresponds to single phonon scattering and can be splitted into two first parts, corresponding to the creation and annihilation of a phonon:

$$\begin{aligned} H_G(\zeta) &= \delta(\zeta)|T_G^{\mathrm{el}}(\zeta)|^2 \\ &\quad + \sum_l \left[\delta(\zeta - \xi_l) n_l |T_{Gl}^{\mathrm{a}}(\zeta)|^2 + \delta(\zeta + \xi_l)(n_l + 1)|T_{Gl}^{\mathrm{c}}(\zeta)|^2\right], \end{aligned} \tag{2.3.1}$$

where the amplitude T_G^{el}, describing Debye-Waller attenuation of the elastic peak, and those of phonon annihilation T_{Gl}^{a} and creation T_{Gl}^{c} may be written as

$$T_G^{\mathrm{el}} = \int_\sigma \frac{\mathrm{d}\boldsymbol{R}}{\sigma} \exp\{-W(\zeta; \boldsymbol{R}) + \mathrm{i}[-\boldsymbol{R}\Delta\boldsymbol{P} + S(\boldsymbol{R})]\}, \tag{2.3.2}$$

$$T_{Gl}^{\mathrm{a}} = \int_\sigma \frac{\mathrm{d}\boldsymbol{R}}{\sigma} \beta_l \exp\{-W(\zeta; \boldsymbol{R}) + \mathrm{i}[-\boldsymbol{R}\Delta\boldsymbol{P} + S(\boldsymbol{R})]\}, \tag{2.3.3}$$

$$T_{Gl}^{\mathrm{c}} = \int_\sigma \frac{\mathrm{d}\boldsymbol{R}}{\sigma} \beta_l^* \exp\{-W(\zeta; \boldsymbol{R}) + \mathrm{i}[-\boldsymbol{R}\Delta\boldsymbol{P} + S(\boldsymbol{R})]\}. \tag{2.3.4}$$

These expressions may be evaluated as follows. Since W and S are periodic functions of $\boldsymbol{R}$, one can expand them into Fourier series over reciprocal lattice vectors $\boldsymbol{G}$. As the coefficients κ_G of these expansions decrease very fast, the terms corresponding to $\boldsymbol{G} \sim (1,0)$, $(-1,0)$, $(0,1)$ and $(0,-1)$ are to be kept in the exponent, while the exponential of the rest may be expanded up to the first power. The result may be expressed through the Bessel functions after a straightforward algebra.

For example,

$$\begin{aligned} T_{Gm}^{\mathrm{c}} &= \exp(-W_0) \sum_{G'} \boldsymbol{e}_m \boldsymbol{d}_{G'} \Delta_{G'm} J_{n_1 - n_1'}(2|\delta_{10}|) J_{n_2 - n_2'}(2|\delta_{01}|) \\ &\quad \times \exp\{\mathrm{i}(n_1 - n_1')\phi_{10} + \mathrm{i}(n_2 - n_2')\phi_{01}\} \end{aligned} \tag{2.3.5}$$

where $\boldsymbol{G} \sim (n_1, n_2)$, $\boldsymbol{G}' \sim (n_1', n_2')$, W_0 is the average of W over σ, J_n is the Bessel function, $\boldsymbol{d}_G = (-\mathrm{i}\boldsymbol{G}, 2\lambda\delta_{n_1 0}\delta_{n_2 0})$, and $\phi_G = \arg\delta_G - \pi/2$; the quantities

$$\Delta_{Gl} = \Delta_{Gl}^+ + \Delta_{Gl}^-, \qquad \Delta_{Gn}^\pm = \kappa_G B^\pm[V_1(z), \omega_l^\pm + \boldsymbol{G}\boldsymbol{V}_{\mathrm{f,i}}], \tag{2.3.6}$$

$$\delta_G = \delta_G^+ + \delta_G^-, \qquad \delta_G^\pm = \kappa_G B^\pm[V_1(z), \boldsymbol{G}\boldsymbol{V}_{\mathrm{f,i}}] \tag{2.3.7}$$

(it is assumed that $\kappa_0 = 1$ here) may be evaluated by (2.1.6)–(2.1.8). The amplitude T_{Gl}^{a} is given by analogous equations.

Actually only few terms make contribution to (2.3.5) due to the following reasons. First, the coefficients κ_G decrease very fast with G, so it is sufficient to take $l_{1,2} = 0, \pm 1$. Second, $B^\pm[V_1, \Omega]$ has a sharp maximum at $\Omega = 0$ (as (2.1.6)–(2.1.8) show), hence only the resonance terms must be taken into account. The resonance condition reads $\omega_l - (\boldsymbol{Q} - \boldsymbol{G}')\boldsymbol{V}_{\mathrm{f,i}} \approx 0$ and has a clear physical sense.

In the thermal regime the corresponding expressions can be obtained by expanding the exponent in (2.1.2). Replacing the sum over m by the integration over the first Brillouin zone

$$\sum_{l} \to \frac{V}{(2\pi)^3} \sum_{j} \int d\boldsymbol{q},$$

where V is the volume of the crystal, and taking into account δ-functions of quasi-momentum and energy conservation yield (Blinov and Kulginov 1991)

$$D(\zeta) = \exp[-2W(\zeta)] \left\{ \delta(\zeta) + \frac{V}{(2\pi)^3} \sum_{j} \left[\sum_{q_z^+} |\beta_l|^2 n_l \chi(\boldsymbol{q}) \left| \frac{\partial \omega_l}{\partial q_z} \right|^{-1} \right.\right.$$

$$\left.\left. + \sum_{q_z^-} |\beta_l|^2 (n_l + 1) \chi(\boldsymbol{q}) \left| \frac{\partial \omega_l}{\partial q_z} \right|^{-1} \right] \right\}, \qquad (2.3.8)$$

here $q_z^\pm$ are the solutions of equation $\Delta E = \omega(\boldsymbol{q}, j)$ with $\boldsymbol{q} = (\pm \Delta \boldsymbol{P}, q_z)$, and $\chi(\boldsymbol{q})$ is the characteristic function of the phonon spectrum.

2.4 Rotational and Vibrational Resonances

In case of coincidence of some characteristic frequencies in gas–surface scattering problem (those of rotation, vibration, thermal oscillation of surface, and "washboard sound") the resonance situation may take place. As a result, the probabilities of corresponding processes may increase dramatically. For instance, the rotational–vibrational resonances of molecular degrees of freedom are governed by the term

$$V_{\mathrm{RV}}(\boldsymbol{r}, \theta, q) = a_0^{l1} V_1(z) P_l(\cos\theta)(\rho - \rho_0) \qquad (2.4.1)$$

in the expansion (1.1.12).

In the framework of our formalism, the description of this resonant interaction is essentially analogous to that of inelastic diffraction ("washboard"–phonon resonance, see brief discussion in the previous section). The trajectories of translational and rotational motion consist of two elastic branches as well, and the result may be expressed through the Bessel functions of the corresponding classical action increments.

If the anisotropy coefficients are small, only the Bessel functions of the order different by unity are involved into the expression for the transition probability. The effect of deep potential well (at low enough scattering energies) weakens the resonances.

3 Multiphonon Scattering

3.1 Gaussian Approximation to the Dynamic Structural Factor

In the hard wall approximation for the Debye model of the phonon spectrum at high enough surface temperatures (when $n_m \approx k_B T_s/\omega_m$) it is possible to obtain an analytical expressions for the Debye–Waller exponent (2.1.3) and the correlation function (2.1.4). Replacing the summation over phonon spectrum by integration yields

$$W(\Delta P, \Delta E) = C \Delta p_z^2, \tag{3.1.1}$$

$$K(\Delta P, \Delta E; R, t) = C \Delta p_z^2 \frac{\mathrm{Si}\,(\omega_D t + q_D R) + \mathrm{Si}\,(\omega_D t - q_D R)}{q_D R}, \tag{3.1.2}$$

where $C = 3T/(2Mk_B T_D^2)$, T_D is the Debye temperature, $\omega_D = k_B T_D = v q_D$ is the Debye frequency, v is the sound velocity, q_D is the Debye wave number and $\mathrm{Si}\,(x)$ denotes the integral sinus. These expressions are quite complicated and cannot be used for low scattering energies, when the wall is "soft", and at low surface temperatures.

In the classical limit the second derivatives of $K(\zeta, \eta)$ at $\eta = 0$ are large, and only a small vicinity of the origin makes contribution to the integral (2.1.2) (this assumption is equivalent to that of short-ranged spatial and temporal correlations of the surface motion). An asymptotic expression for the scattering kernel in this limit may be obtained by expanding the exponent in series and evaluating the integral over η by the Laplace method (Brako and Newns 1982; Manson 1991):

$$-2W + K \approx \mathrm{i}\tau \overline{E} + \mathrm{i}R\overline{P} - \tau^2 \langle \omega^2 \rangle - 2\tau R \langle \omega Q \rangle - RR : \langle QQ \rangle, \tag{3.1.3}$$

where angle brackets stand for averaging over the phonon spectrum:

$$\langle f(q) \rangle = \sum_l \left(n_l + \frac{1}{2} \right) |\beta_l|^2 f(q), \tag{3.1.4}$$

while $\overline{E}$ and $\overline{P}$ are the average energy and tangential momentum transfers to the surface:

$$\overline{E} = \sum_l |\beta_l|^2 \omega_m, \qquad \overline{P} = \sum_l |\beta_l|^2 Q. \tag{3.1.5}$$

As the whole expression is too bulky to be written here, we choose axis 1 at the surface pane parallel to P_i (with axis 2 orthogonal to it) and neglect $\langle \omega Q_2 \rangle$ and $\langle Q_1 Q_2 \rangle$. Such approximation is valid provided the scattering peak is sharp enough. Then

$$D(\Delta P, \Delta E) = (4\pi)^{-3/2} A^{-1/2} \exp\left(-\frac{B}{4} \right), \tag{3.1.6}$$

where

$$A = \langle Q_2^2 \rangle \left(\langle \omega^2 \rangle \langle Q_1^2 \rangle - \langle \omega Q_1 \rangle^2 \right) > 0 \tag{3.1.7}$$

(as the Cauchy inequality guarantees), and

$$B = \frac{K_1^2}{\langle Q_1^2 \rangle} + \frac{K_2^2}{\langle Q_2^2 \rangle} + \frac{\langle Q_1^2 \rangle \langle Q_2^2 \rangle}{A} \Omega^2. \tag{3.1.8}$$

The values K and Ω are related to the tangential momentum and energy gains:

$$K = \Delta P + \overline{P}, \tag{3.1.9}$$

$$\Omega = \Delta E + \overline{E} - \frac{K\langle\omega Q\rangle}{\langle Q_1^2\rangle}. \tag{3.1.10}$$

It is easy to see from these equations that average energy and tangential momentum transfers are really given by (3.1.5). The scattering probability reaches its maximum when there is one single-phonon transition in each mode with the probability $|\beta_m|^2$ (this is a consequence of the harmonic nature of the crystal).

Scattering kernel with the dynamic structural factor (3.1.6) is properly normalized (if the integral over p_f is evaluated by the Laplace method):

$$\int_{p_{\text{f}z}>0} R(p_\text{f}, p_\text{i})\,\mathrm{d}p_\text{f} = 1. \tag{3.1.11}$$

Moreover, if $T_\text{s} \gg T_\text{D}$, so that $n_l \approx k_\text{B}T_\text{s}/\omega_l$, one has

$$\overline{E} = \frac{\langle\omega^2\rangle}{k_\text{B}T_\text{s}}, \quad \overline{P} = \frac{\langle\omega Q\rangle}{k_\text{B}T_\text{s}}, \tag{3.1.12}$$

and R meets the principle of detailed balance with classical Maxwellian gas distribution function (see, for example, (Ferziger and Kaper 1972)):

$$|p_{\text{i}z}|\exp\left(-\frac{p_\text{i}^2}{2m_g k_\text{B}T}\right) R(p_\text{f}, p_\text{i}) = p_{\text{f}z}\exp\left(-\frac{p_\text{f}^2}{2m_g k_\text{B}T}\right) R(-p_\text{i}, -p_\text{f}). \tag{3.1.13}$$

Another advantage of this model is that expression (3.1.6) involves only three quantities related to the surface characteristics and does not depend on any specific model of the phonon spectrum. The contributions of specific surface phonons and adsorbate vibrational modes are additive:

$$\langle f(q)\rangle = \langle f(q)\rangle_\text{bulk} + \langle f(q)\rangle_\text{surface} + \langle f(q)\rangle_\text{adsorbate}\,. \tag{3.1.14}$$

When approximation (2.1.9) with U = const is used, β_l does not depend on ΔP and $\langle Q\rangle = \langle\omega Q\rangle = \overline{P} = 0$, $\overline{E} = \Delta p_z^2/2M$. If, besides that, the trajectory approximation $p_{\text{f}z} \approx -p_{\text{i}z}$ is applied and Δp_z^2 is evaluated assuming that gas atom collides with only one surface atom, then

$$\overline{E} = \frac{4m_g M}{(m_g + M)^2}\left(\frac{p_{\text{i}z}^2}{2m_g} + U\right) \tag{3.1.15}$$

and one obtains Baule formula for the energy loss. Finally, when only surface phonon modes are taken into consideration, $\langle Q_i Q_j\rangle = \delta_{ij}v^2\langle\omega^2\rangle/2$ and one gets Brako–Newns formula (Brako and Newns 1982).

In case when either $V_\text{f,i} \ll v$, or β depends on frequency weakly enough to replace $\beta(\omega_l - QV_\text{f,i})$ with its expansion up to the second power of $V_\text{f,i}$, the average momentum transfer $\overline{P}$ may be expressed through $\overline{E}$. Assuming that the first Brillouin zone is bounded by a sphere or a circle (for surface phonons), integrating by parts over ω gives

$$\overline{P} = \frac{V_\text{i} + V_\text{f}}{2v^2}\left[\frac{\nu+1}{\nu}\overline{E} - \omega_\text{D} N \sum_j \beta_j^2(\omega_\text{D})\right], \tag{3.1.16}$$

where ν is the dimension of $\boldsymbol{q}$: $\nu = 2$ or 3. If β does not depend on frequency (in case of fast collisions), $\overline{E} = 0$ as noted above. When $p_D \ll p_{\mathrm{f,iz}} \ll p^{\pm}(\omega_{\mathrm{D}})$, the second term in parentheses may be neglected and (3.1.16) becomes particularly simple.

3.2 Multiphonon Scattering from Monolayer Adsorbate

Scattering of atoms from the adsorbed rare gas monolayers is a very convenient for theoretical analysis phenomenon. The pair potentials for rare gas–rare gas interactions are well known from gas phase experiments, and the potential of adatom–adsorbed layer interaction may be calculated as a sum of these pair potentials (Chung et al. 1985, 1986). Processes involving electron degrees of freedom do not play significant role in this case.

Experiments show that impinging atoms can excite only the modes corresponding to normal displacements of adsorbed atoms. These modes may interact with surface modes of metal substrate (for example in Kr/Pt(111) system, see (Kern et al. 1987)), but for Ar, Kr, and Xe adsorbed at Ag(111) they are dispersionless (Gibson and Sibener 1988). Their wave vectors are two dimensional. These features are very convenient for further considerations and calculations.

Finally, these layers have low corrugation (it is valid for Ar and Kr at least), so one can neglect diffraction and assume the potential surfaces to be plane. All this, together with accurate experimental results (Gibson and Sibener 1988), makes these systems ideal for the comparison of various approaches and approximations and allows one to consider them as the gauge systems for the verification of different methods of surface probing.

Since the corrugation of adsorbed layer may be neglected, the scattering probability is given by (2.1.1) with the dynamic structural factor (2.1.2). It is convenient to use the phonon expansion of the dynamic structural factor, because one can identify individual processes of phonons creation and annihilation in experiments (Gibson and Sibener 1988; Moses et al. 1992).

The expressions for the probability of scattering with p phonons created and q ones annihilated may be obtained by expanding e^K in powers of K:

$$\begin{aligned} D(\zeta) &= \exp[-2W(\zeta)] \sum_{r=0}^{\infty} \int \frac{\mathrm{d}\eta}{(2\pi)^3} \exp(\mathrm{i}\zeta\eta) K^r(\zeta,\eta) \\ &= \exp[-2W(\zeta)] \sum_{p,q=0}^{\infty} \frac{(n+1)^p n^q}{p!\,q!} D_{pq}(\zeta), \end{aligned} \tag{3.2.1}$$

where

$$\begin{aligned} D_{pq} &= \int \frac{\mathrm{d}\eta}{(2\pi)^3} \underbrace{\sum_{l,l',\dots}}_{p+q} |\beta_l|^2 |\beta_{l'}|^2 \dots \exp[\mathrm{i}\eta(\zeta + \underbrace{\xi_l + \xi_{l'} + \dots}_{p} - \underbrace{\xi_{l''} - \dots}_{q})] \\ &= \delta(\varDelta E + (p-q)\omega) \underbrace{\sum_{l,l',\dots}}_{p+q} |\beta_l|^2 |\beta_{l'}|^2 \dots \delta(\varDelta \boldsymbol{P} + \underbrace{\boldsymbol{Q} + \boldsymbol{Q}' + \dots}_{p} - \underbrace{\boldsymbol{Q}'' - \dots}_{q}). \end{aligned} \tag{3.2.2}$$

In this section n_l and ω_l do not depend on l, hence the subscripts are omitted. The physical sense of these expressions is obvious: term D_{pq} corresponds to p creations and q annihilations of phonons with the probabilities $(n+1)|\beta_l|^2$ and $n|\beta_l|^2$, respectively. It is divided by $p!\,q!$ because phonons are indistinguishable. Different terms do not interfere because the initial state of the adsorbed layer is mixed one, see discussion of this question in (Lipkin 1973).

Replacing the summation over the surface Brillouin zone of the adsorbate (SBZ) by integration

$$\sum_l |\beta_l|^2 \to \frac{\sigma}{4\pi^2}\int_{\mathrm{SBZ}} \mathrm{d}\boldsymbol{Q}\, B(\boldsymbol{Q}), \tag{3.2.3}$$

where $\sigma = a^2\sqrt{3}/2$ is the area of the elementary cell of hexagonal plane lattice with the constant a,

$$B(\boldsymbol{Q}) = N|\beta(\boldsymbol{Q})|^2, \tag{3.2.4}$$

and N is the number of atoms in the adsorbed layer, yields

$$D_{pq} = \delta(\Delta E + (p-q)\omega)\left(\frac{\sigma}{4\pi^2}\right)^{p+q}\widetilde{D}_{pq}. \tag{3.2.5}$$

For $q \neq 0$

$$\widetilde{D}_{pq} = \int_{\mathrm{SBZ}} \mathrm{d}\boldsymbol{Q}_1 \ldots \mathrm{d}\boldsymbol{Q}_p \mathrm{d}\boldsymbol{Q}'_1 \ldots \mathrm{d}\boldsymbol{Q}'_{q-1}\, B(\boldsymbol{Q}_1)\ldots B(\boldsymbol{Q}'_{q-1})B(\boldsymbol{Q}'_q)\,\chi(\boldsymbol{Q}'_q) \tag{3.2.6}$$

with $\boldsymbol{Q}'_q = \Delta\boldsymbol{P} + \boldsymbol{Q}_1 + \ldots + \boldsymbol{Q}_p - \boldsymbol{Q}'_1 - \boldsymbol{Q}'_{q-1}$ and $\chi(\boldsymbol{Q})$ being the characteristic function of the surface Brillouin zone:

$$\chi(\boldsymbol{Q}) = \begin{cases} 1, & \boldsymbol{Q} \in \mathrm{SBZ}, \\ 0, & \text{otherwise}. \end{cases} \tag{3.2.7}$$

If $q = 0$, the expression (3.2.6) should be modified:

$$\widetilde{D}_{p0} = \int_{\mathrm{SBZ}} \mathrm{d}\boldsymbol{Q}_1 \ldots \mathrm{d}\boldsymbol{Q}_{p-1}\, B(\boldsymbol{Q}_1)\ldots B(\boldsymbol{Q}_{p-1})B(\boldsymbol{Q}_p)\,\chi(\boldsymbol{Q}_p) \tag{3.2.8}$$

with $\boldsymbol{Q}_p = -\Delta\boldsymbol{P} - \boldsymbol{Q}_1 - \ldots - \boldsymbol{Q}_{p-1}$.

The probability of projectile scattering with the energy gain $k\omega$ is given by the sum

$$P_k(\Delta\boldsymbol{P}) = \exp[-2W(\Delta\boldsymbol{P}, k\omega)] \sum_{\substack{p,q\geq 0\\ q-p=k}} \left(\frac{\sigma}{4\pi^2}\right)^{p+q} \frac{(n+1)^p n^q}{p!\,q!} \widetilde{D}_{pq}(\Delta\boldsymbol{P}, k\omega). \tag{3.2.9}$$

At low temperature $k_{\mathrm{B}}T_s \lesssim \omega$, n is small and these series converge very fast, so one has to take into account only several first terms.

This expression may be used for calculations directly, but it is useful to discuss its qualitative features. In order to estimate $\widetilde{D}_{pq}$ let us replace the hexagonal surface Brillouin zone by a square with the side length $2Q_M$:

$$4Q_M^2 \frac{\sigma}{4\pi^2} = 1, \qquad Q_M = \frac{\pi\sqrt{2}}{a\sqrt[4]{3}}. \tag{3.2.10}$$

In this case $\chi(\boldsymbol{Q}) = \chi(Q_x/Q_M)\,\chi(Q_y/Q_M)$, where now

$$\chi(x) = \begin{cases} 1, & |x| < 1, \\ 0, & \text{otherwise.} \end{cases} \tag{3.2.11}$$

If, besides that, only the in-plane scattering is considered, $B(\boldsymbol{Q})$ depends only on the parallel component of $\boldsymbol{Q}$, and $\widetilde{D}_{pq}$ may be factorized:

$$\widetilde{D}_{pq} = Q_M^{p+q-1} S_{p+q-1}(0) H_{pq}, \tag{3.2.12}$$

where

$$H_{pq} = \int_{-Q_M}^{Q_M} \mathrm{d}Q_1 \ldots \mathrm{d}Q_p \mathrm{d}Q'_1 \ldots \mathrm{d}Q'_{q-1}\, B(Q_1) \ldots B(Q'_q)\, \chi\left(\frac{Q'_q}{Q_M}\right), \tag{3.2.13}$$

and

$$S_r(y) = \int_{-1}^{1} \mathrm{d}x_1 \ldots \mathrm{d}x_r\, \chi(y + x_1 + \ldots + x_r). \tag{3.2.14}$$

A rough approximation may be obtained by assuming that B depends on Q weakly: $B(Q) \approx B$. This assumption is valid, for example, for impulsive collisions. Then

$$H_{pq} = B^{p+q} (Q_M)^{p+q-1} S_{p+q-1} \left(\frac{\Delta P}{Q_M}\right). \tag{3.2.15}$$

Function $S_r(y)$ has a clear geomerical sense: this is a volume of 2 units thick layer, normal to the main diagonal of the r-dimensional unit cube and shifted along it by y. It is evident that $S_r(y) = 0$ when $|y| > r + 1$.

For large r an asymptotic expression for $S_r(y)$ may be obtained:

$$S_r(y) \underset{r\to\infty}{\sim} 2^r \sqrt{\frac{3}{2\pi r}} \int_{y-1}^{y+1} \exp\left(-\frac{3z^2}{2r}\right) \mathrm{d}z \tag{3.2.16}$$

$$\approx 2^r \sqrt{\frac{6}{\pi r}} \exp\left(-\frac{3y^2}{2r}\right). \tag{3.2.17}$$

So one can see that the probability (3.2.9) is given by the sum of Gaussian-like functions of ΔP of fast decreasing height and slowly increasing width.

4 Scattering from Rough Surfaces

4.1 Statistical Structural Factor

The semiclassical representation (1.4.20) allows one to study scattering from rough surfaces as well. For the sake of simplicity let us limit ourselves with the model of local heights and neglect surface vibrations. Within this model the interaction potential may be presented as $V_s(\boldsymbol{r}) = V_0\big(z - h(\boldsymbol{R})\big)$ with function $h(\boldsymbol{R})$ describing the surface relief. The classic action (1.2.7) may be then approximated as

$$S[\boldsymbol{r}(t)] = -\int_{\boldsymbol{p}_i}^{\boldsymbol{p}_f} \left(\boldsymbol{r} - \frac{\boldsymbol{p}}{m_g} t\right) \mathrm{d}\boldsymbol{p} - \int_{-\infty}^{+\infty} V_0\big(z - h(\boldsymbol{R})\big)\, \mathrm{d}t$$

$$\approx S_0[\boldsymbol{r}(t)] - h(\boldsymbol{R}_0) \Delta p_z, \tag{4.1.1}$$

where S_0 is the action at homogeneous surface, and it is assumed that the surface is locally plane at a height $h(\boldsymbol{R}_0)$, $\boldsymbol{R}_0 = \boldsymbol{R}|_{t=0}$ during a collision event. Assuming that the influence functional $F = 1$ and performing transformations analogous to those leading to (2.2.1)–(2.2.3) yield

$$H_G(\Delta\boldsymbol{P} - \boldsymbol{G}, \Delta E) = \delta(\Delta E)|T_{\Delta P}|^2 \frac{\sigma^2}{4\pi^2 \Sigma} \sum_{k,k'} \exp\{\mathrm{i}[(\boldsymbol{a}_{k'} - \boldsymbol{a}_k)\Delta\boldsymbol{P} + (h_{k'} - h_k)]\}, \tag{4.1.2}$$

with $h_k = h(\boldsymbol{a}_k)$.

Since the exact relief function is never known, one should average (4.1.2) over an ensemble of rough surfaces (this averaging is designated by angle brackets). Furthermore, if the statistical properties of the surface are not changed by translations (the stationary random process model), one of the summations in (4.1.2) may be carried out expicitly:

$$H_G(\Delta\boldsymbol{P} - \boldsymbol{G}, \Delta E) = |T_{\Delta P}|^2 \delta(\Delta E) \frac{\sigma}{4\pi^2} \sum_{k} \exp(\mathrm{i}\boldsymbol{a}_k \Delta\boldsymbol{P}) \langle \exp[\mathrm{i}(h_k - h_0)\Delta p_z] \rangle. \tag{4.1.3}$$

Further consideration needs a model for the random variable h_k to evaluate the correlator. The simplest and widely spread model is the Gaussian random process. In this case the correlator may be expressed through the pair correlation function $\langle h_k h_0 \rangle$

$$\langle \exp[\mathrm{i}(h_k - h_0)\Delta p_z] \rangle = \exp[-\langle h^2 \rangle \Delta p_z] \exp[\langle h_k h_0 \rangle \Delta p_z] \tag{4.1.4}$$

(here $\langle h^2 \rangle = \langle h_k h_k \rangle$ is the mean square height).

When the roughness is small, the coherent component of scattered particles may be separated by expanding $\exp[\mathrm{i}\langle h_k h_0 \rangle \Delta p_z]$ in powers. Up to the second order this expansion reads

$$H_G(\Delta\boldsymbol{P} - \boldsymbol{G}, \Delta E) = \mathrm{e}^{-\langle h^2 \rangle \Delta p_z} |T_{\Delta P}|^2 \delta(\Delta E) \left[1 + \Delta p_z \frac{\sigma}{4\pi^2} \sum_k \mathrm{e}^{\mathrm{i}\boldsymbol{a}_k \Delta\boldsymbol{P}} \langle h_k h_0 \rangle \right], \tag{4.1.5}$$

where $\mathrm{e}^{-\langle h^2 \rangle \Delta p_z}$ is an analog of the Debye–Waller factor, the first term corresponds to the coherent scattering, and the rest does to the single-encounter scattering. All the dropped terms correspond to the multiple-encounter scattering.

The statictical structural factor of the single-encounter scattering is proportional to the Fourier transform of the height–height correlator. Methods of its evaluation have been considered in detail in the light scattering theory. The only essential difference is in the wavelength, the latter being much smaller for a particle beam scattering. This allows one to investigate surface roughness on atomic scale, when this theory is applied for the interpretation of experimental data.

4.2 Influence of Local Defects

Local defects of crystal structure, such as adatoms, dislocations, etc., may be taken into consideration within the framework of the same formalism. For simplicity it is assumed that all the defects have the same nature. The corresponding term in the interaction potential may then be written as $\sum_n V_{\text{def}}(|\boldsymbol{r} - \boldsymbol{R}_n|)$, where $\boldsymbol{R}_n$ is the position of nth defect, and V_{def} is the potential of an individual defect. If V_{def} does not affect the classical trajectory, then the classical action (1.2.7) may be written as

$$S[\boldsymbol{r}] = S_0(\boldsymbol{R}_0) + \sum_n S_{\text{def}}(\boldsymbol{R}_n - \boldsymbol{R}_0), \tag{4.2.1}$$

$$S_{\text{def}}(\boldsymbol{R}) = -\int V_{\text{def}}(|\boldsymbol{r}(t) - \boldsymbol{R}|)\,\mathrm{d}t. \tag{4.2.2}$$

We will assume, for the sake of simplicity, that defects are immobile and surface is "cold". Then, the expression (1.4.20) (where term F, describing surface phonons must be omitted) can be taken as a starting point for the further analysis. Now the scattering kernel for the surface with local defects, averaged over the ensemble of defect positions (denoted by angle brackets), reads

$$\begin{aligned} R(\boldsymbol{p}_\mathrm{f}, \boldsymbol{p}_\mathrm{i}) = \frac{p_{\mathrm{f}z}}{m_g}\delta(\Delta E) \int \frac{\mathrm{d}\boldsymbol{R}_0\,\mathrm{d}\boldsymbol{R}_0'}{4\pi^2 \Sigma} \exp[-\mathrm{i}(\boldsymbol{R}_0 - \boldsymbol{R}_0')\Delta \boldsymbol{P}] \\ \times \left\langle \exp\left\{\mathrm{i}S\left[\boldsymbol{R}_0\right] - \mathrm{i}S\left[\boldsymbol{R}_0'\right]\right\}\right\rangle . \end{aligned} \tag{4.2.3}$$

Our next step will be the derivation of the expansion over the defect encounter multiplicity. For this purpose let us present the exponential factor $\exp[\mathrm{i}\sum_n S_{\text{def}}(\boldsymbol{R}_n - \boldsymbol{R}_0)]$ as a finite expansion in series over $\zeta_n(\boldsymbol{R}_0) = \exp[\mathrm{i}S_{\text{def}}(\boldsymbol{R}_n - \boldsymbol{R}_0)] - 1$

$$\exp[\mathrm{i}\sum_n S_{\text{def}}(\boldsymbol{R}_n - \boldsymbol{R}_0)] = 1 - \sum_n \zeta_n(\boldsymbol{R}_0) + \sum_{n_1 \neq n_2} \zeta_{n_1}(\boldsymbol{R}_0)\zeta_{n_2}(\boldsymbol{R}_0') - \ldots \tag{4.2.4}$$

It is easy to see that quantity ζ_n vanishes, provided projectile does not hit n-th defect. Thus, the k-th term in the expansion (4.2.4) presents scattering event, involving a particle collision with k different defects. Expansion of similar nature was used by other authors (Armand 1987), (Jonsson 1984) in the framework of another formalism. Being inserted in (4.2.3) this expansion leads to

$$\begin{aligned} R(\boldsymbol{p}_\mathrm{f}, \boldsymbol{p}_\mathrm{i}) = \frac{p_{\mathrm{f}z}}{m_g}\delta(\Delta E) \int \frac{\mathrm{d}\boldsymbol{R}_0\,\mathrm{d}\boldsymbol{R}_0'}{4\pi^2 \Sigma} \exp[-\mathrm{i}(\boldsymbol{R}_0 - \boldsymbol{R}_0')\Delta \boldsymbol{P}] \exp[\mathrm{i}S_0(\boldsymbol{R}_0) - \mathrm{i}S_0(\boldsymbol{R}_0')] \\ \times \left\langle 1 - \sum_n \left\{\zeta_n(\boldsymbol{R}_0) + \zeta_n^*(\boldsymbol{R}_0')\right\} + \sum_{n,n'} \zeta_n(\boldsymbol{R}_0)\zeta_{n'}^*(\boldsymbol{R}_0') - \ldots \right\rangle \end{aligned} \tag{4.2.5}$$

We will concentrate our analysis on the contribution of the third term in the angle brackets (it can be shown, that first two terms do not contribute in incoherent part of the scattering). First of all, it is convenient to shift dummy integration variables for every term in the sum over n, n' $\boldsymbol{R}_0 \to \boldsymbol{R}_0 + \boldsymbol{R}_n$, $\boldsymbol{R}_0' \to \boldsymbol{R}_0' + \boldsymbol{R}_{n'}$, so that dynamical part can be factorized from statistical one

$$R^{[3]}(p_f,p_i) = R_{def}(p_f,p_i)\left\langle \frac{1}{N}\sum_{n,n'} \exp\{i(R_n - R_{n'})\Delta P\}\right\rangle \qquad (4.2.6)$$

$$R_{def}(p_f,p_i) = \frac{p_{fz}}{m_g}\delta(\Delta E)\int \frac{dR_0\,dR_0'}{4\pi^2\sigma}\exp[-i(R_0 - R_0')\Delta P]$$
$$\times \exp\{iS_0(R_0) - iS_0(R_0')\}\,\zeta_0(R_0)\zeta_0^*(R_0'), \qquad (4.2.7)$$

with R_{def} being the kernel of scattering from defect.

In order to study the statistical factor, let us first of all introduce the characteristic function of defect positions $X(a_k)$ that is equal to 1, if one of the defects is centered at cell a_k, and 0 othervise. Now we can extend the summation under angle brackets to all surface cells.

$$\left\langle \frac{1}{N}\sum_{n,n'} \exp\{-i(R_n - R_n')\Delta P\}\right\rangle$$
$$= \frac{1}{N}\sum_{k,k'}\langle X(a_k)X(a_{k'})\rangle \exp\{-i(a_k - a_k')\Delta P\}$$
$$= \sum_k \langle X(a_k)X(0)\rangle \exp\{-ia_k\Delta P\} \qquad (4.2.8)$$

The last equality is valid since the defect statistics is assummed to be uniform over the surface. We proved, that scattering kernel have the familiar form of the product of the defect structural factor and the Fourier transform of defects autocorrelator. One can see, that the term in the last sum with $a_k = 0$ gives contribution of order of $\eta = N_{def}/N$, that can be interpreted as a incoherent scattering from defected part of the surface, while other terms present correlation of different defects and have the order of η^2.

The presented approach based on the semiclassical representation of Van Hove type allows one to study inelastic scattering from rough surfaces, as well as the effects of co-existing different scales roughness and superlattices.

5 Concluding Remarks on Quasiclassical Approach to Gas-Surface Scattering

The formalism presented above is important from many points of view. First of all, it is an approach, that can be supposed to be exact, if one knows in detail the interaction potential and the state of the collision system over which the averaging is done. That means, that one deals with a rigorous dynamical description of the collision system that can be a good starting point for different dynamical approximations and derivation of simplified models of collisions for realistic systems. It was shown elsewhere (see i.e. (Bogdanov 1991)), that this formalism was equivalent to the exact close coupling approach and could be used as a source for many well known dynamical approximation, widely used in gas - surface scattering, to mention here only different versions of perturbation theory and sudden approximation just a few.

Of course, such approaches and approximations could be, and actually were obtained by more direct ways. Nevertheless it is very important to have a unified approach at least by three reasons. We can easily establish the range of validity of any dynamical approximation. We can find corrections to pertinent approximation by combining it with some more elucidated approach. We can use some dynamical relations from one approximations and combine them with additional ones from other approaches to improve the analytical properties of the method. Practical computations show, that it is very important to support as many analytical properties of the exact formalism as possible.

Two most impressive examples of the proposed combination of different approaches are the unitarization of the eikonal amplitude, that made possible not only to increase the range of validity of the eikonal approximation, but also to incorporate resonance states into the formalism, and the adiabatic scaling relations, that were obtained by combining the scaling relations from the sudden approximation with adiabatic formulae for the excitation of the ground state.

But the most natural approximations embraced by the proposed approach are the approximations of quasiclassical type. They all can be formulated in terms of a path integral by computation it with different variants of the stationary phase method. Path integral formulation here is not the only one, either but its two important advantages – it is easy to calculate the quasiclassical phase in passing singular points, and it is a straightforward procedure to obtain a uniform approximation, valid also near singularity, – make it without rival with any other approaches.

Not all the possibilities of the quasiclassical approach were used in our discussion. We limited ourselves to so called direct mechanisms of scattering. They are very illustrative for the introduction of the formalism and make it possible to formulate different approximations in terms of so called canonical perturbation theory for the classical action. The defect of such description is obvious. We did not take into account resonance scattering and adsorption-desorption processes. Of course, it is possible to incorporate them into the path integral approach but it will make all the formulae very complicated and shade the physical essence of all the expressions. Therefore we did not give corresponding description. Besides that there are more traditional ways to take such processes into account.

In most practical applications, discussed in this book, one dynamical approximation is used – so called generalized eikonal approach. It is very refined and effective combination of the uniform quasiclassical representation of the scattering amplitude and the eikonal approximation for propagator, that seems to be very natural compromise between the complexity of the problem and our ability to understand its main features and tendencies.

Suggested approach of course can be generalized and advanced. The generalization can be realized in two directions – taking into account next terms of the perturbation theory and incorporaing more complicated trajectories (incontinuous and complex ones). Model examples show how different effects of scattering can be described within such an approach. Nevertheless, it is very important that generalized eikonal formalism gives us an ansatz for the scattering probability in which more refined approaches change the values of coefficients only. That is why it is so convenient for interpretation of experimental data.

II

Microscopic Models of Detailed Kinetics of Adsorption

Introduction

Microscopic models of detailed adsorption kinetics allow one to study non–isothermal effects in the theory of adsorption–desorption, diffusion, surface reactions, and nucleation phenomena, to investigate the influence of adsorbate on gas particles scattering from the surface, to develop methods for kinetic coefficients (sticking, evaporation, diffusion, energy exchange, slipping, etc.) evaluation, and to substantiate phenomenological models widely used in practical calculations of thin film growth dynamics.

It is worth emphasizing that recent publications pay much less attention to the problem of detailed kinetics at interphase boundary in comparison with phenomenological approaches. We explain this fact by the complexity of this problem, whose solution should combine the scattering theory of molecules from inhomogeneous surfaces, Knudsen layer theory, kinetics of non-ideal media, methods of evaluating elementary processes in adsorbate and so on. Another reason is a lack of information on interaction potentials, surface parameters, and incompleteness of experimental data. That is why to our opinion the complete solution of this problem based on the first principles is unrealistic now. Nevertheless, one can develop microscopic models of interphase interaction combining some results of kinetic theory, equilibrium and non-equilibrium thermodynamics, surface physics and surface chemistry.

In what follows we will present some general microscopic models of adsorption kinetics that may be useful before discussing thermodynamical and phenomenological models.

6 Microscopic Models of Detailed Kinetics

We shall deal here with a Lattice Gas (LG) model and a model of Unified Gas–Adsorbate Layer (UGAL). These two models correspond to two alternative approaches in the statistical theory of equilibrium adsorbates (Flood 1967) and can be considered as mutually complementary ones. Which of them should be used depends on the adsorbate properties (ideal, weakly non-ideal, liquid, polycrystalline, localized, partially localized, delocalized), external conditions (isothermal, non-isothermal), mechanisms of elementary processes (adiabatic, non-adiabatic), etc.

Neglecting on the first step the correlations between adsorbed molecules we shall describe the relaxation processes by one-particle distribution function $f_c(\boldsymbol{E}, \boldsymbol{r}, t)$ of molecules of chemical sort c with energy $\boldsymbol{E}$ at point $\boldsymbol{r}$. The adsorbed molecule is deemed to be in the static potential field $V_s(\boldsymbol{r}, \zeta)$ (where ζ is the vector of intramolecular coordinates), that includes both static surface potential and the self-consistent potential of interaction between molecules. Each molecule interacts with the phonon and electron subsystems of solid body (inelastic and non-adiabatic interactions respectively) and collides with other adsorbed molecules. Thus from the point of view of kinetic theory one deals with the relaxation of moving in the field $V_s(\boldsymbol{r},\zeta)$ molecules via the adsorbate–adsorbate collisions, collisions with phonons and electrons. To describe this kinetics the two above mentioned models (LG and UGAL) will be used.

6.1 Lattice Gas (LG) Model

In this model, generalizing the ideas of Dubrovskiy with co-authors (Dubrovskiy and Zyryanov 1987; Dubrovskiy et al. 1988; Dubrovskiy 1991a; 1991b), it is assumed that the potential $V_s(\boldsymbol{r},\zeta)$ determines some fixed space lattice of potential wells centered at discrete points $\boldsymbol{\alpha} = (\beta, \boldsymbol{R})$ ($\beta = 0,1,2,3,\ldots$ is a layer number, $\boldsymbol{R} = n_1\boldsymbol{a}_1 + n_2\boldsymbol{a}_2$, $n_1, n_2 = \pm1, \pm2, \ldots$ is a discrete radius-vector of the plane lattice in β-th layer (see Fig. 1) with $\boldsymbol{a}_1$, $\boldsymbol{a}_2$ being the layer unit cell vectors). Let us assume for the sake of simplicity that a lattice cell $\boldsymbol{\alpha}$ can incorporate only one adsorbed particle. When the latter particle is too big to be incorporated in one cell (adsorption of molecular gases) one should assume instead that for a species c there could be introduced a "cluster" of cells $\boldsymbol{\alpha}_c$ with sufficient room to place one whole molecule of the sort c. A molecule in this case resembles a cluster consisting of separate atoms (Fig. 1). Below $\boldsymbol{E}$ will stand for a set of variables (E, E_{int}), with E being the energy of molecular vibrations in the potential well $\boldsymbol{\alpha}$, and E_{int} being the intramolecular energy ($E = E(l_1, l_2)$; $E_{\text{int}} = E_{\text{int}}(m_1, m_2, \ldots)$; l_1, l_2 are the quantum numbers of vibrational motion, while $m_1, m_2, \ldots$ describe the internal state of the molecule).

In a vicinity of each cell $\boldsymbol{\alpha}$ the potential will be assumed to be of a simplified form

$$V_s(\boldsymbol{r},\zeta) = -\varepsilon_c(\boldsymbol{\alpha}) + V_z(z|\boldsymbol{R}) + V_R(\boldsymbol{R}|\boldsymbol{\alpha}) + V_\zeta(\zeta|\boldsymbol{\alpha}), \tag{6.1.1}$$

where $\varepsilon_c(\boldsymbol{\alpha})$ is the adsorption well depth for a molecule of sort c with the zero energy corresponding to the full energy of a motionless molecule in the ground state for $z \to \infty$; $V_z(z|\boldsymbol{R})$, $V_R(\boldsymbol{R}|\beta)$, $V_\zeta(\zeta|\boldsymbol{\alpha})$ are respectively the profiles of the potential surface in normal and tangential direction, and along variables ζ (right arguments on the right hand side of (6.1.1) represent a weak parametric dependence on corresponding value). It is clear, that this model refers to the localized adsorption. For movable adsorbate and homogeneous layers one can use more simple expression

$$V_s(\boldsymbol{r},\zeta) \approx -\varepsilon_c^{(\beta)} + V_z(z|\beta) + V_\zeta(\zeta|\beta). \tag{6.1.2}$$

For $V_z(z)$ they use one of the empirical potentials with one well, while for $V_R(\boldsymbol{R})$ the approximation of harmonic or anharmonic multidimensional oscillator is generally applied.

Let us introduce the following normalizing condition for the distribution function $f_c(\boldsymbol{E}, \boldsymbol{\alpha}, t)$:

$$\sum_{\boldsymbol{E}} f_c(\boldsymbol{E}, \boldsymbol{\alpha}, t) = \theta_c(\boldsymbol{\alpha}, t), \tag{6.1.3}$$

$$\sum_{c} \theta_c(\boldsymbol{\alpha}, t) = \theta(\boldsymbol{\alpha}, t), \tag{6.1.4}$$

$$0 \le \theta(\boldsymbol{\alpha}, t) \le 1. \tag{6.1.5}$$

In the expressions (6.1.3), (6.1.4) symbol $\sum$ means the sum over discrete and the integral over continuous variables. The overage $\theta(\beta, \boldsymbol{R}, t)$ has the meaning of the

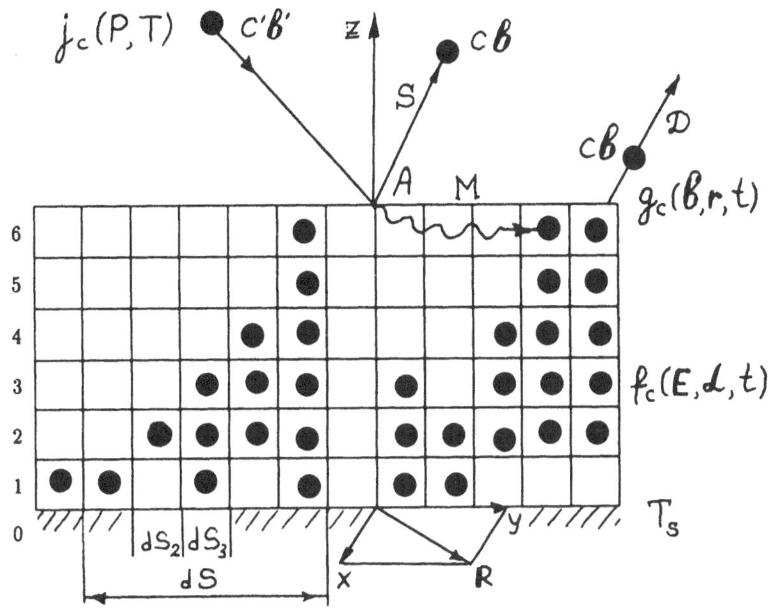

Fig. 1. "Snapshot" configuration of adsorbate on geometrically inhomogeneous substrate and elementary processes at interphase boundary: S – direct scattering, including reactive one; A – adsorption channel; D – desorption channel; M – migration channel; wave line is the sketch of the potential relief for LG model; dS_3, dS_2 – filled areas in layers 3 and 2 of total area dS that are "open" to the gas phase; T, T_s – gas and surface temperatures, respectively.

probability for the cell $(\beta, \boldsymbol{R})$ to be occupied with a molecule at time t. The distribution function $\theta(\beta, \boldsymbol{R}, t)$ gives the distribution density of adsorption microreliefs (see Fig. 1). Along with $\theta(\beta, \boldsymbol{R}, t)$ we shall use quantity $\tilde{\theta}(\beta, \boldsymbol{R}, t)$ defined by

$$\begin{gathered} \tilde{\theta}(\beta, \boldsymbol{R}, t) = Q^{-1}(\boldsymbol{R}, t)\theta(\beta, \boldsymbol{R}, t), \\ Q = \sum_{\beta \geq 0} \theta(\beta, \boldsymbol{R}, t) \\ \theta(0, \boldsymbol{R}, t) = (1 - \theta(1, \boldsymbol{R}, t))(1 - \theta(2, \boldsymbol{R}, t)), \end{gathered} \qquad (6.1.6),$$

that is normalized by the condition

$$\sum_{\beta \geq 0} \tilde{\theta}(\beta, \boldsymbol{R}, t) = 1 \qquad (6.1.7)$$

and has the meaning of the probability for a gas phase molecule to occupy one of the the cells $(\beta, \boldsymbol{R})$ at the point $\boldsymbol{R}$.

In the continuous layers approximation the coverage $\theta(\beta, \boldsymbol{R}, t)$ $(\beta \geq 0)$ can be defined as the ratio of the area $dS_\beta(\boldsymbol{R}, t)$ of a layer β that is "open" for the gas phase to the total area dS (see Fig. 1)

$$\theta(\beta, \boldsymbol{R}, t) = \frac{\mathrm{d}S_\beta(\boldsymbol{R}, t)}{\mathrm{d}S} \tag{6.1.8}$$

In this case the local layer coverage can be calculated as follows

$$\bar{\theta}(\beta, \boldsymbol{R}, t) = \sum_{\beta' \geq \beta} \theta(\beta', \boldsymbol{R}, t) \tag{6.1.9}$$

with the reverse relation having the form

$$\tilde{\theta}(\beta) = \bar{\theta}(\beta) - \bar{\theta}(\beta + 1).$$

The following kinetic parameters of an adsorbed thin film can be expressed through $\tilde{\theta}(\beta, \boldsymbol{R}, t)$: the local adsorbate height $h(\boldsymbol{R}, t)$ in point $\boldsymbol{R}$ at time t, variance of these heights $\Delta(\boldsymbol{R}, t)$, mean film thickness $\bar{h}(t)$ and its variance $\bar{\Delta}(t)$, relative adsorbate volumes $\Omega(\boldsymbol{R}, t) = h(\boldsymbol{R}, t)$ and $\bar{\Omega}(t) = \bar{h}$.

$$\begin{aligned} h(\boldsymbol{R}, t) &= \sum_{\beta \geq 0} \beta \tilde{\theta}(\beta, \boldsymbol{R}, t), \\ \Delta(\boldsymbol{R}, t) &= \sum_{\beta \geq 0} (\beta - h)^2 \tilde{\theta}(\beta, \boldsymbol{R}, t), \\ \bar{h}(t) &= \sum_{\boldsymbol{R}} h(\boldsymbol{R}, t)/S, \\ \bar{\Delta}(t) &= \sum_{\boldsymbol{R}} (h - \bar{h})^2/S. \end{aligned} \tag{6.1.10}$$

Here the quantities h, $\bar{h}$ are measured in units of h_1; Δ, $\bar{\Delta}$ - in units of h_1^2; S is the integral area of the surface under consideration; $h_1 = h_2 = h_3 = \ldots$ are the average heights of the adsorption layers.

The kinetic equation for the distribution function $f_c(\boldsymbol{E}, \boldsymbol{\alpha}, t)$ can be formulated in the following way

$$\partial_t f_c(\boldsymbol{E}, \boldsymbol{\alpha}, t) = \mathsf{E}(f_c) + \mathsf{M}^R(f_c) + \mathsf{M}^\beta(f_c) + \mathsf{AD}(g_c, f_c) + \mathsf{X}(g_c, f_c), \tag{6.1.11}$$

where E, M^R, M^β, AD, X are the kinetic operators of energy relaxation, intralayer migration, interlayer transitions, non-reactive adsorption–desorption, and reactions, respectively. The distribution function $g_c(\boldsymbol{b}, \boldsymbol{r}, t)$ of gas particles is normalized by the following condition

$$\sum_{\boldsymbol{b}} g_c(\boldsymbol{b}, \boldsymbol{r}, t) = n_c(\boldsymbol{r}, t), \qquad \sum_c n_c(\boldsymbol{r}, t) = n(\boldsymbol{r}, t), \;\; \boldsymbol{b} = (\boldsymbol{p}, E_{\mathrm{int}}), \tag{6.1.12}$$

with n_c and n being the partial and total gas particle densities respectively.

The operators E, M^R, M^β, AD will be represented in balance form

$$\mathsf{E}(f_c) = \sum_{\boldsymbol{E}'} [e_c(\boldsymbol{E}', \boldsymbol{E}) f_c(\boldsymbol{E}') - e_c(\boldsymbol{E}, \boldsymbol{E}') f_c(\boldsymbol{E})]; \tag{6.1.13}$$

$$M^R(f_c) = \sum_{\boldsymbol{E}', \boldsymbol{R}'} [m_c(\boldsymbol{E}'\beta\boldsymbol{R}', \boldsymbol{E}\beta\boldsymbol{R}) f_c(\boldsymbol{E}', \boldsymbol{R}') - m_c(\boldsymbol{E}\beta\boldsymbol{R}, \boldsymbol{E}'\beta\boldsymbol{R}') f_c(\boldsymbol{E}, \boldsymbol{R})]; \tag{6.1.14}$$

$$\mathsf{M}^{\beta}(f_c) = \sum_{E',\alpha'} [m_c(E'\alpha', E\alpha) f_c(E'\alpha') - m_c(E\alpha, E'\alpha') f_c(E,\alpha)]\,; \tag{6.1.15}$$

$$\begin{aligned}\mathsf{AD}(g_c, f_c) &= \sum_{b'} [a_c(b', E\alpha) g_c(b') - d_c(E\alpha, b') f_c(E,\alpha)] \\ &= a_{cg}(E\alpha) g_c(b') - d_{c1}(E\alpha) f_c(E\alpha),\end{aligned} \tag{6.1.16}$$

where $e_c(E', E)$, $m_c(E'\alpha', E\alpha)$, $a_c(b', E\alpha)$ and $d_c(E\alpha, b')$ are the probabilities of corresponding events per unit time; $a_{cg}(E\alpha)$ and $d_{c1}(E\alpha)$ are the local adsorption-desorption coefficients.

Surface chemical reactions are assumed to be described by the general formula

$$\sum_{c'=1}^{N_z} z_{c'u} Z_{c'} + \sum_{b'=1}^{N_g} y_{b'u} Y_{b'} \underset{K_u^{\mathrm{r}}}{\overset{K_u^{\mathrm{d}}}{\leftrightarrows}} \sum_{c=1}^{N_z} z_{cu} Z_c + \sum_{b=1}^{N_g} y_{bu} Y_b. \tag{6.1.17}$$

Here $z_{c'u}$, $y_{b'u}$ (z_{cu},y_{bu}) are the stoichiometric coefficients; $Z_{c'}$, $Y_{b'}$ (Z_c, Y_b) are the chemical symbols of the reactants (products) involved in the chemical reaction from adsorbed (Z) and gaseous (Y) phases; index u enumerates reactions resulting in the production of c component; K_u^{d} and K_u^{r} are the detailed rate constants of direct and reverse reactions. Balance equation for the operator X for reactions of the type (6.1.17) reads

$$\begin{aligned}\mathsf{X}(g_c, f_c) = \sum_u \sum_{\delta,\delta'} \Big[& K_u^{\mathrm{d}} \prod_m f_{c'_m}(E'_m, \alpha'_m) \prod_n g_{c'_n}(b'_n) \\ & - K_u^{\mathrm{r}} \prod_l f_{c_l}(E_l, \alpha_l) \prod_k g_{c_k}(b_k) \Big]\end{aligned} \tag{6.1.18}$$

with $K_u^{\mathrm{d}} = K_u^{\mathrm{d}}(\delta', \delta)$ and $K_u^{\mathrm{r}} = K_u^{\mathrm{r}}(\delta, \delta')$, where δ' and δ represent the sets of the relevant variables.

The representation of the kinetic operators in the form (6.1.13)–(6.1.16), (6.1.18) does not reduce essentially the general nature of the kinetic equation (6.1.11) because the non-ideality of the media can be (up to the certain degree of approximation) taken into account in expressions for the transition probabilities. The latter can be determined by the methods of equilibrium and non-equilibrium thermodynamics, with the help of dynamical models for elementary processes, and via empirical formulae.

Along with (6.1.11) it is worth putting down an equation for $\theta_c(\alpha, t)$ that is formally obtained by summing (6.1.11) over E and using (6.1.3)

$$\partial_t \theta_c(\alpha, t) = \mathsf{M}_*^R(\theta_c) + \mathsf{M}_*^{\beta}(\theta_c) + \mathsf{AD}_*(j_c, \theta_c) + \mathsf{X}_*(j_c, \theta_c). \tag{6.1.19}$$

The operators M_*^R, M_*^{β}, AD_*, X_* can still be represented in a balance form similar to (6.1.13)–(6.1.16), (6.1.18) with the following substitutions $f_c \to \theta_c$, $g_c \to j_c$, $m_c \to m_{c*}$, $a_c \to a_{c*}$, $d_c \to d_{c*}$, $K_u^{\mathrm{d,r}} \to K_{u*}^{\mathrm{d,r}}$. The quantities m_{c*}, a_{c*}, d_{c*}, $K_{u*}^{\mathrm{d,r}}$ are the averages of the corresponding non-starred probabilities over the energy variables. In fact, the explicit form of the operators M_*^R, M_*^{β}, AD_*, X_* should be defined through the combined solution of the equations (6.1.11), (6.1.19) by the Enskog–Chapman method, moment method, and so forth (see, e.g., (Kolesnichenko 1986)).

In some cases these operators can be found from independent considerations (e.g. for isothermal processes they can be estimated with the help of the methods of thermodynamical kinetics (see below)).

Using definitions (6.1.10) one can derive from (6.1.19) equations for the film macroprofile $h(\boldsymbol{R},t)$ and its fluctuations $\Delta(\boldsymbol{R},t)$, the film thickness $\bar{h}(t)$ and its variance $\bar{\Delta}(t)$. Equation (6.1.19) will be studied in more detail for the case of a structureless gas under isothermal conditions in what follows.

In conclusion, it is worth emphasizing that the LG model is a natural one for the description of the adsorbed multilayer growth. It is closely connected to the Ising model, which has been intensively discussed in the literature. At last, this model is convenient for studying the scattering of fast particles (electrons, neutrons, molecules) from surfaces with adsorbate.

6.2 Unified Gas-Adsorbate Layer (UGAL) Model

The main idea of the UGAL model, taking its origin from the statistical theory of a fluid adsorbate (Flood 1967), is that a gas phase contacting with a surface, continuously transforms into a layer of adsorbed molecules. The latter evolves in a static field of the substrate $V_s(\boldsymbol{r},\zeta)$ and in a fluctuating field of the surface thermal vibrations. These fields force a molecule c to enter the adsorption well along a stochastic trajectory, analogous to that in the Brownian motion (see Fig. 2.). The calculation of phase diagrams at equilibrium is reduced to including the perturbation of the gas or fluid system by the surface. In nonequilibrium case the problem is in the kinetics of the adsorbate creation through the stochastic mechanism of a molecule capture by the adsorbtion well.

To formulate the UGAL model let us introduce the distribution function of gaseous particles $g_c(\boldsymbol{b},\boldsymbol{r},t)$ normalized by (6.1.12). The description of the adsorbate kinetics is reduced then to solving a Cauchy problem

$$\begin{aligned} \partial_t g_c(\boldsymbol{b},\boldsymbol{r},t) &= \mathsf{J}(g_c), \\ g_c(\boldsymbol{b},\boldsymbol{r},0) &= g_{c0}(\boldsymbol{b},\boldsymbol{r}), \end{aligned} \tag{6.2.1}$$

where $\mathsf{J}(g_c)$ is the operator of the relaxation of molecules over variables $\boldsymbol{b}$ and c. For $z > z_*$ (z_* is the molecule–surface interaction radius) $\mathsf{J}(g_c)$ transforms into the kinetic operator for the gas phase adjacent to the surface, while for $0 \le z < z_*$ it is the operator of the molecule relaxation in the adsorbed phase. The latter can be calculated within various approximations developed in the kinetic theory of non-ideal media. For example, for structureless non-ideal gases the following kinetic equation of quasiparticle type has been obtained (Dubrovskiy and Bogdanov 1979b)

$$\mathsf{J}(g) = -\{g,\varepsilon\} + \mathsf{I}(g), \tag{6.2.2}$$

$$\varepsilon = \frac{p^2}{2\mu} + V(\boldsymbol{r}) + \int \Gamma(\boldsymbol{p},\boldsymbol{p}')g(\boldsymbol{p}')\mathrm{d}\boldsymbol{p}'. \tag{6.2.3}$$

Here $\{g,\varepsilon\}$ is the classic Poisson bracket, ε is the total energy with the account of self–consistent field, $\mathsf{I}(g)$ is the collision integral of Boltzmann–Landau type, the kernel $\Gamma(\boldsymbol{p},\boldsymbol{p}')$ and the collision cross-section in $\mathsf{I}(g)$ are expressed through the amplitude of binary quasiparticle scattering.

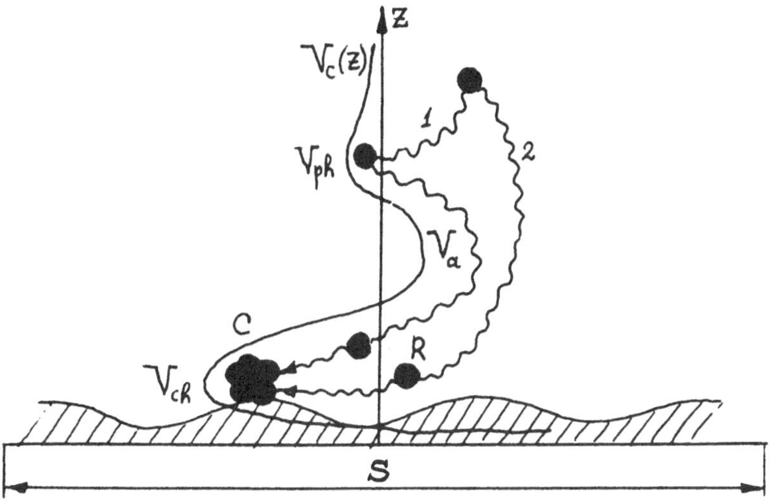

Fig. 2. Elementary processes at interphase boundary for mobile adsorbate. $V_c(z)$ - averaged potential of molecule–surface interaction; V_{ch}, V_{ph} - chemisorption and phisisorption well depths; V_a - chemisorption barrier; 1, 2 - optional paths of molecule diffusion in the potential well; C - cluster; R - reactive interaction of molecules.

In the kinetics of non-ideal media the relaxation operator $\mathsf{J}(g_c)$ is often split into two parts, describing "strong" and "weak" interactions

$$\mathsf{J}(g_c) = \mathsf{F}(g_c) + \mathsf{N}(g_c) \tag{6.2.4}$$

Operator $\mathsf{F}(g_c)$, responsible for the "weak" interactions, has the Fokker-Planck form, while "strong" interactions operator $\mathsf{N}(g_c)$ is approximated in different ways, depending on the type and mechanism of the transition. The latter operator is responsible for large momentum transfers and significant changes of physical and chemical characteristics of the molecule. For example, in case of inelastic interactions and chemical reactions one can use representation (6.1.18) with the only difference, that in the UGAL model there stands a unified distribution function $g_c(\boldsymbol{b}, \boldsymbol{r}, t)$. For large momentum transfer the model of hard spheres is used.

If "weak" interactions lead only to small momentum increments one can approximate the operator $\mathsf{F}(g_c)$ as

$$\mathsf{F}(g_c) = \begin{cases} -\boldsymbol{p}/\mu \partial_{\boldsymbol{r}} g_c + (\partial_{\boldsymbol{r}} V_c) \partial_{\boldsymbol{p}} g_c + \sum_{ij} \partial_{p_i} [D^{p}_{ij} (p_j/\mu + k_{\mathrm{B}} T \partial_{p_j})] g_c, & t_p \ll t_r; \\ \sum_{ij} \partial_{p_i} [D^{p}_{cij} (p_j/\mu + k_{\mathrm{B}} T \partial_{p_j})] g_c \\ \qquad + \partial_{r_i} [D^{r}_{cij} (\partial_{r_j} + \frac{1}{k_{\mathrm{B}} T} (\partial_{r_j} V_c))] g_c, & t_p \approx t_r. \end{cases} \tag{6.2.5}$$

In these relations t_p, t_r are the characteristic times of molecule relaxation over the variables $\boldsymbol{p}$ and $\boldsymbol{r}$, V_c is the self-consistent field potential for the molecule c, D^{p}_{cij},

D^r_{cij} – are the tensors of molecular diffusion in subspaces $\boldsymbol{p}$ and $\boldsymbol{r}$. For isotropic case one has

$$D^p_{cij} = \delta_{ij} D^p_c; \qquad D^r_{cij} = \delta_{ij} D^r_c. \tag{6.2.6}$$

Diffusion coefficients D^p_c, D^r_c for the Brownian model of motion are related via the Einstein's formula (Akhiezer and Peletminskiy 1977)

$$D^r_c D^p_c = k_{\rm B} T/m. \tag{6.2.7}$$

For the density $n_c(\boldsymbol{r}, t)$ of molecules of sort c one can get an equation of Smoluchowski type

$$\begin{aligned} \partial_t n_c(\boldsymbol{r}, t) &= \partial_{\boldsymbol{r}} [D^r_{c*}(\partial_{\boldsymbol{r}} + \frac{1}{k_{\rm B}T} \partial_{\boldsymbol{r}} V_c)] n_c(\boldsymbol{r}, t) + \mathsf{X}_*(n_c), \\ n_c(\boldsymbol{r}, 0) &= n_{c0}(\boldsymbol{r}), \end{aligned} \tag{6.2.8}$$

that is valid for $t_p \ll t_r$ in the approximation of "strong friction", and for $t_p \approx t_r$ as well. In (6.2.8) D^r_{c*} is the averaged over the energy spatial diffusion coefficient, while $\mathsf{X}_*(n_c)$ is the averaged over the energy reactive interaction operator. From mathematical point of view (6.2.8) represents a typical example of nonlinear equation of "Reaction + Diffusion" type intensively discussed in the literature (see, for example, (Kudryavtsev 1987)).

Effects of the non-ideality of adsorbate are incorporated here through the introduction of a dependence of potential V_c, diffusion coefficient D^r_{c*}, and rate constants of chemical reactions in the operator X_* on the distribution function g_c. These dependencies can be found from dynamical models of elementary processes, statistical thermodynamics of equilibrium and nonequilibrium processes, and from experimental data (see, e.g., (Croxton 1974)).

From the above said it follows that the UGAL model is convenient for the description of mechanisms of film growth from gaseous and liquid phase in isothermal conditions, when the temperatures and molecular free paths are small, "trajectories" of molecules have stochastic character, and the adiabatic mechanisms of elementary processes take place.

6.3 Knudsen Layer Problem with Adsorbate

Gaseous media near a surface is perturbed by the scattering and adsorption–desorption processes, as well as the surface chemical reactions (see Figs. 1,2). This perturbation may be taken into account via a kinetic boundary condition (KBC) at $z = z_* \gtrsim \bar{h}$ (in the scale of the gas phase kinetic parameters the value z_* can be set to zero) for the distribution function g_c. The kinetic boundary problem (Knudsen layer problem) for the distribution function $g_c(\boldsymbol{b}, \boldsymbol{r}, t)$ is formulated in the following way

$$\partial_t g_c(\boldsymbol{b}, \boldsymbol{r}, t) + \boldsymbol{v} \partial_{\boldsymbol{r}} g_c(\boldsymbol{b}, \boldsymbol{r}, t) = \mathsf{J}(g_c); \tag{6.3.1}$$

$$p_z g^+_c(\boldsymbol{b}) = \sum_{b'c'} R_{c'c}(\boldsymbol{b}', \boldsymbol{b}) g^-_{c'}(\boldsymbol{b}') |p'_z| + D_{cf}(\boldsymbol{b}); \quad g_c{}^{\pm} = g_c(\boldsymbol{b})\big|_{p_z \gtrless 0}; \tag{6.3.2}$$

where $R_{cc'}(\boldsymbol{b}', \boldsymbol{b})$ is the probability of direct inelastic scattering (channel S in Fig. 1), $D_{cf}(\boldsymbol{b})$ is the probability of desorption (channel D in Fig. 1). Note that the KBC

(6.3.2) is inhomogeneous because particles scattered directly (including resonant scattering through bound state) and the particles desorbing from the surface after relaxation in adsorbed layer are treated in different ways. Since the probabilities $R_{c'c}$, D_{cf} depend on f_c, the KBC (6.3.2) describes the influence of relaxational phenomena in the adsorbate on the gas phase perturbation at the surface. Different forms of KBC are discussed in Part IV.

When calculating adsorption-desorption probabilities one has to use models, that allow one to couple the energetic variables in the gas phase with those in the adsorbed layer. Let us divide the translational motion of an adsorbed molecule into normal and tangential components. We shall denote the energies of the molecular motion along x, y, z axes as E_x, E_y, E_z respectively. The momentum of the projectile molecule in the same coordinate system can be presented as

$$p' = (p_x', p_y', p_z'); \quad p'_k = \sqrt{2m_c E'_k}; \quad k = x, y, z. \tag{6.3.3}$$

Adsorbed particle states are characterized by negative energies E_k. Then non-reactive adsorption-desorption processes may be treated as the transitions between the energetic states of translational motion

$$E_k' \leftrightarrows E_k, \quad E_k' > 0, \quad E_k < 0. \tag{6.3.4}$$

To simplify the calculations they used to suppose that all the processes occur in one plane.

The probabilities $R_{c'c}(b', b)$ and $D_{cf}(b)$ in the LG model are functionals of the distribution functions f_c and g_c (see Part IV)

$$R_{c'c} = R_{c'c}(b', b | g_c, f_c), \qquad D_{cf} = D_{cf}(b | f_c). \tag{6.3.5}$$

Taking into account the dependence of the right hand side of Eq. (6.1.11) on g_c and that of $R_{c'c}$, $D_{c'c}$ on f_c, allows one to conclude that the problems of adsorbed and Knudsen layers are closely connected. This connection may be important for the problems of the film growth and for the problems of gas–surface interaction as well (e.g. for the calculation of the slip and exchange coefficients, heat and mass exchange at the boundary, and so on).

One can evaluate the kinetic boundary operators in (6.3.2) on the base of the UGAL model as well. Such results for some simple interaction and kinetics models have been published elsewhere (Cercignani 1975; Balakhonov and Zak 1988).

Having at one's disposal kinetic equations for adsorbate and gas phase one has an opportunity to consider different regimes and derive for these regimes appropriate macroscopic transport equations by the kinetic theory methods. The investigations of transport processes at interphase boundary in the framework of UGAL - type models have been presented elsewhere (Borisov et al. 1988; Borman et al. 1988).

7 Physical Models of Elementary Processes, Transition Probabilities, and Kinetic Coefficients

To describe the detailed kinetics at interphase boundary within the LG or UGAL models one has to know a large number of the probabilities and rate constants of elementary and kinetic processes. Depending on the information available they can be taken from classical and quantum dynamical models, thermodynamics and phenomenology, and from experiment. We are going to consider here some models of elementary processes in adsorbed layer and corresponding approximations for the probabilities and kinetic coefficients.

7.1 Model of 3D LG Adsorbate

Vibrational spectra of adsorbed molecules have been studied in a lot of publications (see, e.g., the list of references in (Zhdanov 1988)). These data can be used for the modeling of two- or one-dimensional adsorption wells. The information on the probabilities of energy relaxation through the interaction with phonon and electron subsystems of solid body is also available from the literature (Zhdanov 1988).

If an adsorbed molecule is moving in electronically adiabatic potential well the model of two-dimensional anharmonic oscillator can be used for the description of its vibrational spectrum. This spectrum $E(n)$ $(n = (n_1, n_2))$ can be calculated by the numerical or analytical integration of the two-dimensional Schrödinger equation. The calculations of the transition probabilities $e(E', E)$ (or $e(n', n)$) for such oscillator have been performed with the help of the perturbation theory or more sophisticated approaches. Provided these transition probabilities $e(n', n)$ are known for dense enough energy spectrum ($\Delta E \ll k_B T_s$), a diffusion model of energy relaxation may be used

$$\mathsf{E}(f_c) = \sum_{j,k=1}^{2} \rho_j(E)\partial_E \left[\rho_k(E) D^n_{jk} \left(\partial_E + \frac{1}{k_B T_s}\right)\right] f_c, \tag{7.1.1}$$

$$D^n_{jk}(E) = \sum_{n'} \frac{(n'_j - n_j)(n'_k - n_k)}{2} e(n', n);$$

$$\rho_i(E) = \partial_{n_i} E(n), \qquad i = j, k$$

with D^n_{jk} being the tensor of energy diffusion in the quantum number space.

For isotropic case (the model of independent oscillators) one has

$$D^n_{jk}(E) = \delta_{jk} D^n_j(E_j); \quad \rho_j(E) = \rho_j(E_j); \quad E_j = E_j(n_j), \tag{7.1.2}$$

$$E_j = E_j(n_j), \quad j = 1, 2.$$

with E_j being the quantum levels of independent vibrations in normal and tangential directions. For independent harmonic oscillators

$$D^E_j = c_j E_j, \quad c_j = \text{const}, \quad \rho_j = \text{const}. \tag{7.1.3}$$

Within the approximation (7.1.2) the operator $\mathsf{E}(f_c)$ becomes

$$\mathsf{E}(f_c) = \sum_{j=1}^{2} \rho_j(E_j)\partial_{E_j} \left[\rho_j(E_j) D_j^n(E_j)(\partial_{E_j} + \frac{1}{k_B T_s}) \right] f_c. \qquad (7.1.4)$$

To include the intramolecular structure one has to consider the interaction of the intramolecular (rotational, vibrational, electronic) degrees of freedom with the translational motions (i.e. vibrations in the adsorption well).

To take into account non-adiabatic effects induced by the avoided-crossing of different potential curves (see Fig. 3), one has to include strong interaction of several vibrational modes of the adsorbate.

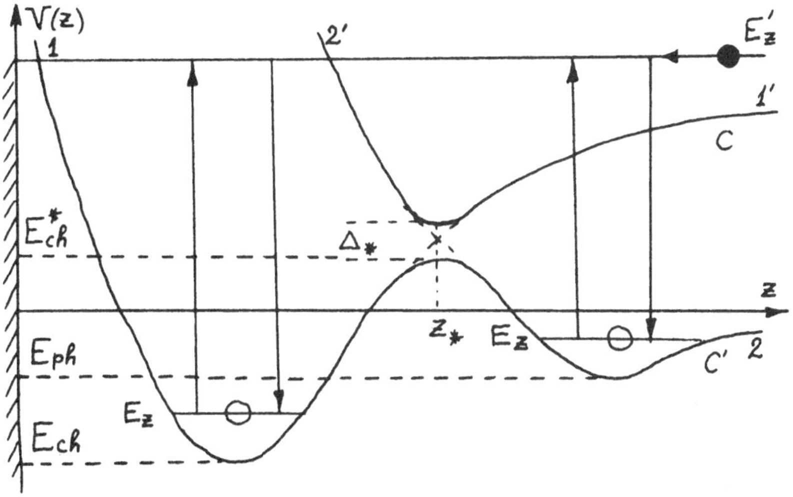

Fig. 3. Potential curves, describing adiabatic and non-adiabatic adsorption mechanisms, 22′, 11′ – diabatic terms, responsible for the configurations $c' + S'$ and $c + S$, respectively for $z \to \infty$ (c and S being adsorbed molecule of the sort c and surface); 21, 2′1′ – adiabatic terms responsible for the same configurations for $z \to \infty$; $E_{\rm ch}$, $E_{\rm ph}$, $E^*_{\rm ch}$ – energies of chemisorption, physisorption and chemisorption activation; Δ_* – nonadiabatic terms interaction parameter; z_* – term crossing point; E'_z, E_z – normal energies of translational motion of molecule in initial and final states. Term 1 corresponds to the chemisorptional state while the term 2 – to the phisisorptional one for $z < z_*$.

In all the above mentioned cases the calculation of spectrum and transition probabilities is a complicated problem of multidimensional (and multimode) vibrational spectroscopy that has just begun to be studied.

For the interlayer transition probabilities in the model of chemisorption with precursor (Dubrovskiy and Zyryanov 1987) one can use a simple representation (with $\beta, \beta' = 1, 2, 3$ standing for the chemisorption, physisorption, and gas phase layers respectively)

$$m(\beta' \boldsymbol{E}' \to \beta \boldsymbol{E}) = \delta(\boldsymbol{E}' - \boldsymbol{E}) \nu_{1\beta'}(E'_z) P_{\beta'\beta}(E'_z, E_z) W_{\beta'\beta}(E'_z, E_z), \qquad (7.1.5)$$

where E_z is the normal motion energy , E – the tangential motion energy (the cylindrical symmetry is now assumed), $\nu_{1\beta'}$ is the frequency of particle collisions with the potential wall dividing layers β' and β, $P_{\beta'\beta}(E'_z, E_z)$ is the probability of a transition from layer β' to β via electronic mechanism, and $W_{\beta'\beta}(E'_z, E_z)$ is that due to energy transfer $\Delta E_z = E'_z - E_z$ to the solid phonon subsystem. In the frame of non–adiabatic two–level model of chemisorption (see Fig. 3) the following expression for $P_{\beta'\beta}(E'_z, E_z)$ has been obtained ($\beta = 1$, $\beta' = 2, 3$) (Bogdanov et al 1985a):

$$P_{\beta'1}(E'_z, E_z) = \left[1 - \exp\left(-\frac{\Delta_*}{E'_z - E^*_{\text{ch}}}\right)\right] \delta(E'_z - E_z)\chi(E'_z - E^*_{\text{ch}}), \qquad (7.1.6)$$

where Δ^* is the parameter of non–adiabatic interaction of crossing terms, E^*_{ch} is the chemisorption activation energy, $\delta(y)$ and $\chi(y)$ are delta–function and Heaviside functions, respectively. For $\Delta^* \gg 1$ and $\Delta^* \ll 1$ one has adiabatic and non–adiabatic limits for the chemisorption probability. For the probability $W_{3\beta}(E'_z, E_z)$ one can use the following expression (Fedotov 1988):

$$W_{3\beta}(E'_z, E_z) = \frac{1}{2\sqrt{\pi(b_1 + b_2)k_B T_s}} \exp\left(-\frac{(E'_z - E_z - b_1)^2}{4k_B T_s(b_1 + b_2)}\right), \qquad (7.1.7)$$

where parameters b_1 and b_2 are proportional to the matrix elements of molecule interaction with the phonon subsystem, the former having a sense of mean energy transfer ($b_1 = \langle E_z \rangle$) while the latter describing the molecule internal degrees of freedom involvement into energy exchange processes. Other approximations for $W_{3\beta}$ can be derived using the methods of Part I.

For a two-layer adsorbate let us suppose that for the first two layers ($\beta = 1, 2$) formed mostly due to the substrate–molecule interaction (Fig. 3), the above considerations are still applicable and molecule may occupy physisorption cells ($\beta = 2$) over both filled (external precursor) and free (internal precursor) chemisorption cells ($\beta = 1$). Therefore, simple approximations for the detailed probabilities of the molecule capture to the adsorption well given by (7.1.6), (7.1.7) are valid. Recently the equilibrium properties of such two–layer adsorbate within the model of chemisorption via precursor have been studied on a base of the lattice gas thermodynamical Hamiltonian (Kreuzer 1990). The obtained results constitute a basis for studying nonequilibrium properties of two-layer adsorbates by the methods of thermodynamical kinetics (see Chapts.9,10).

In case of multilayer adsorbate ($\beta > 2$) certain hypotheses on the capture mechanism and interaction potential of colliding particle with the adsorbate are necessary for the calculation of the adsorption probabilities. Particularly, in some cases the expressions (7.1.6) and (7.1.7) are valid. Omitting the question of the calculation of detailed adsorption-desorption probabilities let us formulate a model based on ideas of the two–layer adsorption model and BET–approximation for the adsorption isotherms of multilayer films (Flood 1967). The main assumptions of this generalized kinetic BET-model are:

1. The first two layers are filled in a way described in the preceding paragraph;
2. For all the rest layers we postulate the absence of "hanging" molecules, i.e. a molecule can occupy the cell ($\beta, \boldsymbol{R}$) only if the underlying cell ($\beta - 1, \boldsymbol{R}$) is filled;
3. The interlayer transitions are allowed to take place;

4. The lateral interactions are included.

The assumptions (1–3) can be taken into account by the following representation of the elementary transition probabilities in (6.1.14)–(6.1.16) (Dubrovskiy 1991b):

$$m_c(\boldsymbol{E}'\boldsymbol{\alpha}', \boldsymbol{E}\boldsymbol{\alpha}) = q(\beta,\theta)\tilde{m}_c(\boldsymbol{E}'\boldsymbol{\alpha}', \boldsymbol{E}\boldsymbol{\alpha}), \tag{7.1.8}$$

$$a_c(\boldsymbol{b}', \boldsymbol{E}\boldsymbol{\alpha}) = q(\beta,\theta)\tilde{a}_c(\boldsymbol{b}', \boldsymbol{E}\boldsymbol{\alpha}),$$

$$q(\beta,\theta) = \begin{cases} [1-\theta(\beta\boldsymbol{R})], & \beta = 1,2 \\ [1-\theta(\beta\boldsymbol{R})]\theta(\beta-1,\boldsymbol{R}), & \beta \geq 3 \end{cases}. \tag{7.1.9}$$

For the averaged over energy transition probabilities m_*, a_* in (6.1.19) the following relations are valid within the generalized BET–model

$$m_{c*}(\boldsymbol{\alpha}', \boldsymbol{\alpha}) = q(\beta,\theta)\tilde{m}_{c*}(\boldsymbol{\alpha}', \boldsymbol{\alpha}), \tag{7.1.10}$$

$$a_{c*}(\boldsymbol{\alpha}) = q(\beta,\theta)\tilde{a}_{c*}(\boldsymbol{\alpha}).$$

At last, for the detailed rate constants $K_{\mathrm{u}}^{\mathrm{d,r}}$ in (6.1.18), (6.1.19) analogous representations may be used with an eye to the nature of the adsorption cells

$$K_u(\boldsymbol{\delta}', \boldsymbol{\delta}) = q_u\tilde{K}_u(\boldsymbol{\delta}', \boldsymbol{\delta}), \tag{7.1.11}$$

$$K_{u*}(\boldsymbol{\delta}', \boldsymbol{\delta}) = q_u'\tilde{K}_{u*}(\boldsymbol{\delta}', \boldsymbol{\delta}).$$

The form of the functions q_u, q_u' in (7.1.11) is more complicated compared to $q(\beta,\theta)$ in (7.1.10) and depends on the geometry of the incorporation of molecules–reagents (products) in initial (final) state into the adsorption cells. The explicit form of these functions can be determined on the base of more detailed information on reaction processes (6.1.17).

From the general expression (6.1.14) one can obtain the following representation for the operator M in diffusion approximation

$$\mathsf{M}(f_c) \approx \sum_{i,j} \partial_{R_i}\left[D^R_{cij}(E,\boldsymbol{R})\left(\partial_{R_j} - K_{ci}(E\boldsymbol{R})\right)\right] f_c; \tag{7.1.12}$$

$$D^R_{cij} \equiv \frac{1}{2}\sum_{\Delta_E,\zeta} \zeta_i\zeta_j\Delta_E^2 m_c(E\boldsymbol{R}, E+\Delta_E\,\boldsymbol{R}+\zeta),$$

$$K_{ci} = \sum_{\Delta_E,\zeta} \zeta_i\Delta_E m_c(E\boldsymbol{R}, E+\Delta_E\,\boldsymbol{R}+\zeta). \tag{7.1.13}$$

In isotropic case more simple expressions for the kinetic coefficients may be used

$$D^R_{cij} = \delta_{ij}D^R_c(E),$$

$$K_{ci} = -(\partial_{R_i}V_c)/k_{\mathrm{B}}T_s; \tag{7.1.14}$$

$$D^R_c(E) = B_c(E)\exp\left(-\frac{E_{cD}}{k_{\mathrm{B}}T_s}\right). \tag{7.1.15}$$

In (7.1.15) E_{cD} is the diffusion activation barrier and B_c – the pre-exponential factor depending on the tangential energy. The latter has been calculated within the theory of correlation functions (Doll and Voter 1985) and the transition state approach (Voter and Doll 1984) with the account of tunnel effects (Zhdanov 1985).

By analogy with (6.2.8) we can use the diffusion approximation for the operator $\mathsf{M}_*(\theta_c)$ instead of discrete variant as well

$$\mathsf{M}_*(\theta_c) = \partial_R \left[D^R_{c*}(\partial_R + \frac{1}{k_B T_s}(\partial_R V_{c*})) \right] \theta_c. \quad (7.1.16)$$

Note, that the Fokker-Planck operator in (7.1.16) differs from that in (6.2.8) since in the latter case it describes the three-dimensional space diffusion, while in the former one – the diffusion over the layer β.

The introduction of factor q in (7.1.8), (7.1.10) is equivalent to including the repulsive part of the pair potential (Langmuir correction) and configurations without inner "cavities" in upper layers. Including the lateral interaction in more detail (the attractive part of the binary potential in particular) leads to the appearance of $\tilde{m}_c$, $\tilde{a}_c$ and $\tilde{m}_{c*}$, $\tilde{a}_{c*}$ dependence on the distribution function θ_c.

7.2 UGAL Model

This model, as was discussed in Chap.6, gives one an opportunity to describe the kinetics of non–ideal gas media in static and fluctuating surface field. Therefore, when approximating the kinetic operators (6.2.4), (6.2.5) one can use the results of quasiparticle method for non–ideal media kinetics (Dubrovskiy and Bogdanov 1979b), theory of liquids (Croxton 1974), theory of Brownian motion (Akhiezer and Peletminskiy 1977), theory of phase transitions, models of equilibrium properties of such systems (Jaycock and Parfitt 1981) with further application of methods of statistical thermodynamics of irreversible processes (Kreuzer and Payne 1988b) and experimental data on pair correlation function (Flood 1967).

Thus, the UGAL model gives additional possibilities for studying the adsorption kinetics of gaseous and liquid film formation.

7.3 Mechanisms of Surface Reactions

Let us consider now some important mechanisms of surface reactions, described by formulae (6.1.17) (we shall focus our attention on bimolecular reactions).

First mechanism states that during a collision of two molecules surface plays the role of a "third body" that can absorb part of the energy and momentum, decrease the activation barrier, perturb the reaction path or capture one of the colliding molecules into a bound state prior to their direct interaction. The possibilities listed correspond to the following channels of a binary surface reaction (M', M'_1; M, M_1 denote molecules in initial and final states in the gas phase)

$$M' + M'_1 + S \leftrightarrows M + M_1 + S \qquad \text{(impact mechanism)}, \quad (7.3.1)$$

$$M' + M'_1 + 2S \leftrightarrows M' + (M'_1 S) + S \leftrightarrows M + M_1 + 2S \quad \text{(Eley – Rideal mechanism)}. \quad (7.3.2)$$

The second mechanism includes the interaction of the electronic subsystems of molecule and solid body leading to the transition of "molecule + surface" system from one potential curve (reagents) to another one (products) via a non–adiabatic (non–adiabatic mechanism) or adiabatic (adiabatic mechanism) transition (see Fig. 3). These reactions can be described by the following scheme

$$M' + M_1' + 2S \leftrightarrows (M'S)^* + (M_1'S)^* \leftrightarrows M + M_1 + 2S, \qquad (7.3.3)$$

where $(M'S)^*$, $(M_1'S)^*$ denote intermediate quasistationary complexes, that correspond to the system on the potential curve decaying to final reaction channel.

For example, systems $A + S$, $A^+ + S$ and $A_2 + 2S$ are usually described by the curve 22′ in Fig. 3, while those $A^- + S^+$, $A + S^+, 2A + 2S$ - by the curve 11′. That is why reactions $A + S \leftrightarrows A^- + S^+$ (surface ionization), $A^+ + S \leftrightarrows A + S^+$ (surface neutralization), $2A + 2S \leftrightarrows A_2 + 2S$ (surface recombination) are assumed to go through non-adiabatic mechanism (Kreuzer and Payne 1988a; 1988b; 1989; Brivio and Grimley 1979; Nørskov and Stoltze 1987; Scott 1980). The first two reactions are important for studying electromechanical phenomena at surfaces (Zandberg and Ionov 1969; Wetter 1967), while the third one - for studying heat transfer to the vehicle surface in atomic oxygen and nitrogen flows (Nørskov and Stoltze 1987; Scott 1980).

The third mechanism of surface reactions involves the adsorption of both molecules - reagents, their diffusion along the surface, two-dimensional "collision", reaction at surface, and desorption of the products (Langmuir–Hinshelwood mechanism):

$$M' + M_1' + 2S \leftrightarrows (M'S) + (M_1'S) \leftrightarrows (MS) + (M_1S) \leftrightarrows M + M_1 + 2S. \qquad (7.3.4)$$

Note that a classification of the surface reaction mechanisms can be done either on the base of the nature of limiting stages, or on the base of dynamical models of elementary acts. The first way of classification is conditional, depending strongly on the relative values of different terms in the equations of chemical kinetics (6.1.19) or (6.3.1). Classification on the base of dynamical models (non-adiabatic, adiabatic, collinear, impact, stochastic, etc.) needs the detailed study of the physical nature of reactive interactions. Such study is at the very beginning now both in theoretical and experimental (molecular beams) directions, and it should lead to detailed information on mechanisms of surface reactions.

8 Different Regimes of Adsorption Kinetics

8.1 Dimensionless Form of Kinetic Equations

Let us provide an analysis of possible regimes of relaxation on the base of equations (6.1.11), (6.1.19). First of all, the operators AD, AD$_*$ with the help of (6.1.16), (7.1.8), (7.1.10) can be rewritten as follows

$$\begin{aligned} \mathsf{AD}(g_c, f_c) &= a_{cg}(\boldsymbol{E}\alpha) - d_{c1}(\boldsymbol{E}\alpha) f_c, \\ \mathsf{AD}_*(j_c, \theta_c) &= a_{cg*}(\boldsymbol{E}\alpha) - d_{c1*}(\alpha)\theta_c(\alpha, t), \\ a_{cg} &= q(\beta, \theta)\tilde{a}_{cg}(\boldsymbol{E}\alpha), \\ a_{cg*} &= q(\beta, \theta)\tilde{a}_{cg}(\boldsymbol{E}\alpha). \end{aligned} \qquad (8.1.1)$$

In dimensionless form the equations (6.1.11), (6.1.19) take on the form

$$\partial_t f_c(\boldsymbol{E},\boldsymbol{\alpha},t) = \varepsilon_E^{-1}\mathsf{E}(f_c) + \varepsilon_R^{-1}\mathsf{M}^R(f_c) + \varepsilon_\beta^{-1}\mathsf{M}^\beta(f_c) + \varepsilon_A^{-1}\eta^{-1}a_{cg}(g_c, f_c) - \varepsilon_D^{-1}d_{c1}f_c + \varepsilon_X^{-1}\mathsf{X}(g_c, f_c) \equiv \mathsf{U}(f_c, g_c); \tag{8.1.2}$$

$$\partial_t\theta_c(\boldsymbol{\alpha},t) = \varepsilon_R^{-1}\mathsf{M}_*^R(\theta_c) + \varepsilon_\beta^{-1}\mathsf{M}_*^\beta(\theta_c) + \varepsilon_A^{-1}\eta^{-1}a_{cg*}(j_c,\theta_c) - \varepsilon_D^{-1}d_{c1*}\theta_c + \varepsilon_X^{-1}\mathsf{X}_*(j_c,\theta_c) \equiv \mathsf{U}_*(\theta_c, j_c). \tag{8.1.3}$$

In (8.1.2), (8.1.3) the following dimensionless parameters have been introduced

$$\varepsilon_l = t_l/t_0; \quad l = E, R, \beta, A, D, X, \qquad \eta = T_s/T, \tag{8.1.4}$$

where t_l is the relaxation time of a molecule over the variable l, t_0 is the characteristic time of the distribution function f_c variation. Thus, different regimes of adsorption kinetics correspond to different relations between the relaxation parameters ε_l and the value of the parameter η of thermal non-equilibrium. The most important regimes of adsorbate relaxation will be considered in detail in what follows.

8.2. Fast Relaxation

In this case all relaxation times t_l are small compared to t_0 and the thermal equilibrium takes place, that is

$$\varepsilon \equiv \varepsilon_l \ll 1, \quad l = E, R, \beta, A, D, X; \qquad \eta = 1, \tag{8.2.1}$$

The kinetic equation (8.1.2) acquires then the following singularly perturbed form

$$\partial_t f_c(\boldsymbol{E},\boldsymbol{\alpha},t) = \varepsilon^{-1}\mathsf{U}(f_c, g_c), \tag{8.2.2}$$

$$\mathsf{U}(f_c, g_c) = \mathsf{E}(f_c) + \mathsf{M}^R(f_c) + \mathsf{M}^\beta(f_c) + \mathsf{AD}(f_c, g_c) + \mathsf{X}(f_c, g_c).$$

If Knudsen number ε_g for the gas phase is small, one comes to the following singularly perturbed problem for g_c

$$\partial_t g_c(\boldsymbol{b},\boldsymbol{r},t) + \boldsymbol{v}\partial_{\boldsymbol{r}} g_c(\boldsymbol{b},\boldsymbol{r},t) = \varepsilon_g^{-1}\mathsf{J}(g_c), \quad \varepsilon_g \ll 1; \tag{8.2.3}$$

$$p_z g_c^+(\boldsymbol{b}) = \sum_{\boldsymbol{b}',c'} R_{cc'}(\boldsymbol{b}',\boldsymbol{b}|f_c, g_c)|p_z'|g_{c'}^-. \tag{8.2.4}$$

Different forms of the kernel $R_{cc'}(\boldsymbol{b}',\boldsymbol{b}|f_c,g_c)$, that depend functionally on the distribution functions f_c, g_c, are discussed in Part IV.

The condition (8.2.1) allows one to expand the solution of (8.2.2), (8.2.3) in power series over small parameters ε, ε_g

$$f_c = f_c^0 + \varepsilon f_{c1} + \varepsilon^2 f_{c2} + \ldots, \tag{8.2.5}$$

$$g_c = g_{c0} + \varepsilon_g g_{c1} + \varepsilon_g^2 g_{c2} + \ldots. \tag{8.2.6}$$

The zero-order distribution functions are to be found from conditions

$$\mathsf{U}(f_c^0, g_{c0}) = 0, \tag{8.2.7}$$

$$\mathsf{J}(g_{c0}) = 0, \tag{8.2.8}$$

$$p_z g_{c0}^+(\boldsymbol{b}) = \sum_{\boldsymbol{b}'c'} R_{c'c}(\boldsymbol{b}',\boldsymbol{b}|f_c^0, g_{c0})|p_z'|g_{c'0}^-(\boldsymbol{b}'). \tag{8.2.9}$$

Let us suppose, that the solution to (8.2.8), (8.2.9) is a local Maxwell-Boltzmann distribution describing locally equilibrium state of the gas phase.

$$\begin{aligned} g_{c0}(\boldsymbol{b},\boldsymbol{r}) &= n_{co}(\boldsymbol{r})\phi_{co}(\boldsymbol{E})\Big|_{z=0} \\ &= n_{c0}(\boldsymbol{R})Q_{cg}^{-1}\exp\left[-\frac{1}{k_{\mathrm{B}}T}\left(\frac{p^2}{2m_c}+E_{\mathrm{int}}\right)\right] \\ &= n_0 Q_{cg}^{-1}\exp\left[-\frac{1}{k_{\mathrm{B}}T}\left(-\mu_{cg}+E_z+E_\tau+E_{\mathrm{int}}\right)\right] \end{aligned} \qquad (8.2.10)$$

with

$$p^2=p_z^2+p_\tau^2;\quad E_z=\frac{p_z^2}{2m_c}\geq 0;\quad E_\tau=\frac{p_\tau^2}{2m_c}\geq 0;\quad E_{\mathrm{int}}\leq 0;\quad \boldsymbol{E}=(E_1,E_2,E_3).$$

In (8.2.10) Q_c is the statistical sum over the energy states ($Q_c=\sum_{\boldsymbol{E}}\exp[-(E_1+E_2+E_{\mathrm{int}})/k_{\mathrm{B}}T]$), and μ_{cg} is the chemical potential of c-th component in the gas phase. An obvious normalization condition holds

$$\sum_{\boldsymbol{E}}\phi_{co}(\boldsymbol{E})=1. \qquad (8.2.11)$$

To find zero order approximation f_c^0 from the condition (8.2.7) one has to make certain assumptions on the transition probabilities e_c, $\tilde{m}_c$, $\tilde{a}_c$, $\tilde{d}_c$, $\tilde{K}_{\boldsymbol{u}}^{\mathrm{d,r}}$. Let us introduce the probabilities of unit acts of adsorption ($\bar{a}_c$), desorption ($\bar{d}_c$), and direct and reverse reactions (6.1.17) ($\bar{K}_{\boldsymbol{u}}^{\mathrm{d}}$, $\bar{K}_{\boldsymbol{u}}^{\mathrm{r}}$) according to

$$\begin{aligned} \tilde{a}_c(\boldsymbol{b}',\boldsymbol{E\alpha})&=\sigma j_c\bar{a}_c(\boldsymbol{b}',\boldsymbol{E\alpha});\quad \tilde{d}_c(\boldsymbol{E\alpha},\boldsymbol{b}')=\nu_{c1}\bar{d}_c(\boldsymbol{E\alpha},\boldsymbol{b}');\\ \tilde{a}_{cg}(\boldsymbol{E\alpha})&=\sigma j_c\bar{a}_{cg}(\boldsymbol{E\alpha});\quad \tilde{d}_{c1}(\boldsymbol{E\alpha})=\nu_{c1}\bar{d}_{c1}(\boldsymbol{E\alpha}); \qquad (8.2.12)\\ \tilde{K}_{\boldsymbol{u}}^{\mathrm{d}}\equiv\tilde{K}_{\boldsymbol{u}}d(\delta',\delta)&=r'_{\boldsymbol{u}}\bar{K}_{\boldsymbol{u}}^{\mathrm{d}}(\delta',\delta);\quad \tilde{K}_{\boldsymbol{u}}^{\mathrm{r}}(\delta,\delta')=r_{\boldsymbol{u}}\bar{K}_{\boldsymbol{u}}^{\mathrm{r}}(\delta,\delta');\\ j_c=p_c\bar{Z}_c;\quad \bar{Z}_c&=(2\pi m_c k_{\mathrm{B}}T)^{-\frac{1}{2}};\quad r'_{\boldsymbol{u}}=r_{\boldsymbol{u}}(\delta'),\quad r_{\boldsymbol{u}}=r_{\boldsymbol{u}}(\delta). \qquad (8.2.13) \end{aligned}$$

In the expressions (8.2.12), (8.2.13) σ is the area of a cell, p_c is the partial pressure of c-th component, ν_{c1} is the frequency of molecule vibration in normal direction, $r'_{\boldsymbol{u}}$ and $r_{\boldsymbol{u}}$ are the correction factors taking into account the difference between the probability of reaction transition per unit time and the probability of unit-act reactive transition from initial to final configuration.

We shall postulate that the local Maxwell–Boltzmann distribution of adsorbed molecules (lattice gas) has the following form

$$\begin{aligned} f_{c0}(\boldsymbol{E},\boldsymbol{\alpha})&=\phi_{co}(\boldsymbol{E},\boldsymbol{\alpha})\theta_{co}(\boldsymbol{\alpha})\\ &=Q_{ca}^{-1}\exp\left[-\frac{1}{k_{\mathrm{B}}T}\left(\varepsilon_c(\boldsymbol{\alpha})-\mu_{ca}+E_z+E_\tau+E_{\mathrm{int}}\right)\right]; \qquad (8.2.14)\\ Q_{ca}&=\sum_{\boldsymbol{E}}\exp\left[-\frac{1}{k_{\mathrm{B}}T}\left(E_z+E_\tau+E_{\mathrm{int}}\right)\right], \qquad (8.2.15) \end{aligned}$$

where

$$\varepsilon_c(\boldsymbol{\alpha})\leq E_1\leq 0;\quad E_2\leq 0;\quad E_{\mathrm{int}}\leq 0.$$

In these expressions Q_c and μ_{ca} are the statistical sum over energy states and the chemical potential of c-th component in the adsorbate.

The following detailed balance relations will be assumed to be held

$$f_{c0}(\boldsymbol{E}')e_c(\boldsymbol{E}',\boldsymbol{E}) = f_{c0}(\boldsymbol{E})e_c(\boldsymbol{E},\boldsymbol{E}'), \tag{8.2.16}$$

$$f_{c0}(\boldsymbol{E}',\boldsymbol{R}')\tilde{m}_c(\boldsymbol{E}'\boldsymbol{R}',\boldsymbol{E}\boldsymbol{R}) = f_{c0}(\boldsymbol{E},\boldsymbol{R})\tilde{m}_c(\boldsymbol{E}\boldsymbol{R},\boldsymbol{E}'\boldsymbol{R}'), \tag{8.2.17}$$

$$g_{c0}(\boldsymbol{b}')\bar{a}_c(\boldsymbol{b}',\boldsymbol{E}\boldsymbol{\alpha}) = f_{c0}(\boldsymbol{E},\boldsymbol{\alpha})\bar{d}_c(\boldsymbol{E}\boldsymbol{\alpha},\boldsymbol{b}'), \tag{8.2.18}$$

$$\prod_m f_{c'_m0}(\boldsymbol{E}'_m,\boldsymbol{\alpha}'_m)\prod_n g_{c'_n0}(\boldsymbol{b}'_n,\boldsymbol{r}_n)\bar{K}^{\mathrm{d}}_u(\boldsymbol{\delta}',\boldsymbol{\delta}) = \prod_l f_{c_l0}(\boldsymbol{E}_l,\boldsymbol{\alpha}_l)\prod_k g_{c_k0}(\boldsymbol{b}_k,\boldsymbol{r}_k)\bar{K}^{\mathrm{r}}_u(\boldsymbol{\delta},\boldsymbol{\delta}'). \tag{8.2.19}$$

Averaging relations (8.2.17)–(8.2.19) over $\boldsymbol{E}$ yields the detailed balance conditions for the averaged probabilities

$$\exp\left(-\frac{\varepsilon_c(\boldsymbol{\alpha}')}{k_{\mathrm{B}}T}\right)\tilde{m}_{c*}(\boldsymbol{R}',\boldsymbol{R}) = \exp\left(-\frac{\varepsilon_c(\boldsymbol{\alpha})}{k_{\mathrm{B}}T}\right)\tilde{m}_{c*}(\boldsymbol{R},\boldsymbol{R}'), \tag{8.2.20}$$

$$\bar{a}_{c*}(\boldsymbol{\alpha}) = \exp\left(-\frac{\varepsilon_c(\boldsymbol{\alpha})}{k_{\mathrm{B}}T}\right)\bar{d}_{c*}(\boldsymbol{\alpha}), \tag{8.2.21}$$

$$\prod_m \exp\left(-\frac{\varepsilon_{c_m}(\boldsymbol{\alpha}')}{k_{\mathrm{B}}T}\right)\prod_n n_{c'_n0}(\boldsymbol{r}_n)\bar{K}^{\mathrm{d}}_{u*}(\boldsymbol{\delta}',\boldsymbol{\delta}) = \prod_l \exp\left(-\frac{\varepsilon_{c_l}(\boldsymbol{\alpha})}{k_{\mathrm{B}}T}\right)\prod_k n_{c_k0}(\boldsymbol{r}_k)\bar{K}^{\mathrm{r}}_{u*}(\boldsymbol{\delta},\boldsymbol{\delta}'). \tag{8.2.22}$$

It is important to emphasize that the averaged detailed balance relations (8.2.20)–(8.2.22) have more wide range of validity than (8.2.17)–(8.2.19) since the adsorbate is a thermodynamical (not fully determined) system. Therefore the further considerations will be restricted to the evaluation of quasiequilibrium distributions $\theta^0_c(\boldsymbol{\alpha})$ vanishing operators M^R_*, M^β_*, AD_*, and X_*.

One can easily prove that the distribution function $f^0_c(\boldsymbol{E},\boldsymbol{\alpha})$ may be presented as

$$f^0_c(\boldsymbol{E},\boldsymbol{\alpha}) = \phi_{c0}(\boldsymbol{E},\boldsymbol{\alpha})\theta^0_c(\boldsymbol{\alpha}). \tag{8.2.23}$$

where $\theta^0_c(\boldsymbol{\alpha})$ obeys the following conditions

$$\frac{\theta^0_c(\boldsymbol{R}'\beta)}{q'(\theta^0)}\exp\left(\frac{\varepsilon_c(\boldsymbol{R}'\beta)}{k_{\mathrm{B}}T}\right) = \frac{\theta^0_c(\boldsymbol{R}\beta)}{q(\theta^0)}\exp\left(\frac{\varepsilon_c(\boldsymbol{R}\beta)}{k_{\mathrm{B}}T}\right); \tag{8.2.24}$$

$$\frac{\theta^0_c(\boldsymbol{R}\beta')}{q'(\theta^0)}\exp\left(\frac{\varepsilon_c(\boldsymbol{R}\beta')}{k_{\mathrm{B}}T}\right) = \frac{\theta^0_c(\boldsymbol{R}\beta)}{q(\theta^0)}\exp\left(\frac{\varepsilon_c(\boldsymbol{R}\beta)}{k_{\mathrm{B}}T}\right); \tag{8.2.25}$$

$$\frac{\theta^0_c(\boldsymbol{\alpha})}{q(\theta^0)} = \frac{\sigma j_c}{\nu_{c1}}\exp\left(\frac{\varepsilon_c(\boldsymbol{\alpha})}{k_{\mathrm{B}}T}\right); \tag{8.2.26}$$

$$\frac{r'_u\prod_m \theta^0_{c'_m}(\boldsymbol{\alpha}'_m)}{q'_u}\prod_m \exp\left(\frac{\varepsilon_{c'_m}(\boldsymbol{\alpha}'_m)}{k_{\mathrm{B}}T}\right) = \frac{r_u\prod_l \theta^0_{c_l}(\boldsymbol{\alpha}_l)}{q_u}\prod_l \exp\left(\frac{\varepsilon_{c_l}(\boldsymbol{\alpha}_l)}{k_{\mathrm{B}}T}\right). \tag{8.2.27}$$

The problem of solving these equations will be considered in the next chapter.

8.3 Fast Isothermal Energy Relaxation

In this case one has

$$\eta \approx 1, \quad \varepsilon \equiv \varepsilon_E \ll 1, \quad \varepsilon_l \approx 1, \quad l = R, \beta, A, D, X \tag{8.3.1}$$

and the kinetic equation (8.1.2) takes on a form

$$\partial_t f_c(\boldsymbol{E}, \boldsymbol{\alpha}, t) = \varepsilon^{-1}\mathsf{E}(f_c) + \mathsf{U}_1(f_c, g_c), \tag{8.3.2}$$

$$\mathsf{U}_1(f_c, g_c) = \mathsf{M}^R(f_c) + \mathsf{M}^\beta(f_c) + \mathsf{AD}(f_c, g_c) + \mathsf{X}(f_c, g_c). \tag{8.3.3}$$

The distribution functions f_c, g_c can be expanded as in (8.2.5), (8.2.6). The zero order term of f_c expansion can be presented as

$$f_c^0(\boldsymbol{E}, \boldsymbol{\alpha}) = \phi_{c0}(\boldsymbol{E}, \boldsymbol{\alpha})\theta_c(\boldsymbol{\alpha}, t) \tag{8.3.4}$$

with $\theta_c(\boldsymbol{\alpha}, t)$ satisfying the equation

$$\partial_t \theta_c(\boldsymbol{\alpha}, t) = \mathsf{U}_*(\theta_c, j_c), \tag{8.3.5}$$

$$\mathsf{U}_*(\theta_c, j_c) = \mathsf{M}_*^R(\theta_c) + \mathsf{M}_*^\beta(\theta_c) + \mathsf{AD}_*(\theta_c, j_c) + \mathsf{X}_*(\theta_c, j_c). \tag{8.3.6}$$

The latter equation is obviously that of isothermal concentration kinetics, where the operators M_*, AD_*, and X_* are obtained by averaging operators M, AD, and X over $\boldsymbol{E}$ with the equilibrium distribution function $\phi_{c0}(\boldsymbol{E})$. Equation (8.3.5) is of importance for the thin film growth kinetics.

The following variants of isothermal kinetics are possible, depending on the relations between corresponding relaxation times ($\varepsilon \ll 1$):

$$\partial_t \theta_c(\boldsymbol{\alpha}, t) = \varepsilon^{-1}\left[\mathsf{M}_*^R(\theta_c) + \mathsf{M}_*^\beta(\theta_c)\right] + \mathsf{AD}_*(\theta_c, j_c) + \mathsf{X}_*(\theta_c, j_c) \tag{8.3.7}$$

(the diffusion of particles is fast, adsorption-desorption and reactive processes are slow);

$$\partial_t \theta_c(\boldsymbol{\alpha}, t) = \varepsilon^{-1}\left[\mathsf{M}_*^R(\theta_c) + \mathsf{M}_*^\beta(\theta_c) + \mathsf{AD}_*(\theta_c, j_c)\right] + \mathsf{X}_*(\theta_c, j_c) \tag{8.3.8}$$

(the diffusion of particles and adsorption-desorption are fast, reactive processes are slow);

$$\partial_t \theta_c(\boldsymbol{\alpha}, t) = \varepsilon^{-1}\left[\mathsf{M}_*^R(\theta_c) + \mathsf{M}_*^\beta(\theta_c) + \mathsf{X}_*(\theta_c, j_c)\right] + \mathsf{AD}_*(\theta_c, j_c) \tag{8.3.9}$$

(the diffusion of particles and reactions are fast, adsorption-desorption processes are slow).

The evaluation of the higher order terms of f_c expansion, and the expansion of solutions to (8.3.7) - (8.3.9) over small parameter can be done in a standard way using the theory of singularly perturbed equations (Kolesnichenko 1986).

8.4 Non-Isothermal Relaxation

Provided the following relations hold

$$\eta \neq 1, \quad \varepsilon_E < \varepsilon_l \approx 1, \tag{8.4.1}$$

$f_c(\boldsymbol{E}, \alpha, t)$ can be presented in the form

$$f_c(\boldsymbol{E}, \alpha, t) = \phi_c(\boldsymbol{E}, t)\theta_c(\alpha, t). \tag{8.4.2}$$

Substituting this expression into (8.1.2) and assuming that the distribution function $\theta_c(\alpha, t)$ varies with time slower than $\phi_c(\boldsymbol{E}, t)$, yields a coupled set of equations

$$\begin{aligned} \partial_t\phi_c(\boldsymbol{E}, t) &= \mathsf{E}(\phi_c) + N_+(\boldsymbol{E}) - N_-(\boldsymbol{E})\phi_c(\boldsymbol{E}, t), \\ \partial_t\theta_c(\alpha, t) &= \mathsf{M}_*^R(\theta_c) + \mathsf{M}_*^\beta(\theta_c) + \mathsf{AD}_*(\theta_c, j_c) + \mathsf{X}_*(\theta_c, j_c). \end{aligned} \tag{8.4.3}$$

The first equation here is the energy relaxation equation with positive and negative sources. The quantities N_+, N_- depend functionally on θ_c (we omit here these dependencies). Averaged over the energy operators M_*^R, M_*^β, AD_*, X_* are to be calculated using non-equilibrium distribution function $\phi_c(\boldsymbol{E}, t)$ (nonisothermal concentration kinetics). The representation of the general equation (8.1.2) in the form (8.4.3) allows one to make certain motivated assumptions for obtaining general solution by iteration method.

Using kinetic theory methods on a base of above mentioned kinetic equations one can derive transport equations for the macroparameters that describe different regimes of mass and heat transfer to the substrate and the thin film formation dynamics.

8.5 Diffusion Form of LG Model Equations

In case of movable adsorbate in some applications it is more convenient to present the migration operator M_*^R in the equation (6.1.19) in diffusion approximation. For the sake of simplicity, let us consider this question on the example of structureless particles when the equation (6.1.19) with an eye to (7.1.8)–(7.1.10) and (8.2.12) takes on the form

$$\partial_t\theta(\alpha, t) = \mathsf{M}_*^R(\theta) + \mathsf{L}_*(\theta, j), \tag{8.5.1}$$

$$\mathsf{M}_*^R(\theta) = \sum_{\boldsymbol{R}'} \left[m(\boldsymbol{R}', \boldsymbol{R})q(\theta)\theta(\boldsymbol{R}', t) - m(\boldsymbol{R}, \boldsymbol{R}')q'(\theta)\theta(\boldsymbol{R}, t)\right], \tag{8.5.2}$$

$$\mathsf{L}_*(\theta, j) = \mathsf{M}_*^\beta(\theta) + \mathsf{AD}_*(\theta, j), \tag{8.5.3}$$

$$\mathsf{M}_*^\beta(\theta) = \sum_{\beta'} \left[m(\beta', \beta)q(\theta)\theta(\beta', t) - m(\beta, \beta')q'(\theta)\theta(\beta, t)\right], \tag{8.5.4}$$

$$\mathsf{AD}_*(\theta, j) = j\sigma q(\theta)a(\alpha) - \nu_1 d(\alpha)\theta(\alpha, t), \tag{8.5.5}$$

where

$$\begin{aligned} m(\boldsymbol{R}, \boldsymbol{R}') &\equiv \tilde{m}_*(\beta, \boldsymbol{R}; \beta, \boldsymbol{R}'), \quad m(\beta, \beta') \equiv \tilde{m}_*(\beta, \boldsymbol{R}; \beta', \boldsymbol{R}); \\ a(\alpha) &= \tilde{a}_{g*}(\alpha), \quad d(\alpha) = d_{1*}(\alpha). \end{aligned} \tag{8.5.6}$$

The probabilities $m(\boldsymbol{R},\boldsymbol{R}')$, $m(\beta,\beta')$, $a(\alpha)$ and $d(\alpha)$ satisfy detailed balance principle

$$\exp\left(-\frac{\varepsilon(\boldsymbol{R}')}{k_{\mathrm{B}}T}\right) m(\boldsymbol{R}',\boldsymbol{R}) = \exp\left(-\frac{\varepsilon(\boldsymbol{R})}{k_{\mathrm{B}}T}\right) m(\boldsymbol{R},\boldsymbol{R}'), \tag{8.5.7}$$

$$\exp\left(-\frac{\varepsilon(\beta')}{k_{\mathrm{B}}T}\right) m(\beta',\beta) = \exp\left(-\frac{\varepsilon(\beta)}{k_{\mathrm{B}}T}\right) m(\beta,\beta'), \tag{8.5.8}$$

$$a(\alpha) = \exp\left(-\frac{\varepsilon(\alpha)}{k_{\mathrm{B}}T}\right) d(\alpha), \tag{8.5.9}$$

with $\varepsilon(\boldsymbol{R}) \equiv \varepsilon(\beta,\boldsymbol{R})$, $\varepsilon(\beta) \equiv \varepsilon(\beta,\boldsymbol{R})$. It should be noted that we have neglected in (8.5.4) the variation of $\boldsymbol{R}$ during adatom transition from one layer to another one in the probabilities $m(\beta,\beta')$, distribution function $\theta(\beta,t)$ and in $q(\theta)$, thus taking into account the interlayer jumps only at the boundaries of adsorption layers.

Suppose that the probabilities of jumps $\boldsymbol{R} \to \boldsymbol{R} + \boldsymbol{R}''$ in the operator M_*^R vary strongly with $\boldsymbol{R}''$ compared to more smooth dependence of the quasiequilibrium distribution function $\theta_R(\boldsymbol{R})$ $\theta_R(\boldsymbol{R}) \approx \theta_R(\boldsymbol{R}+\boldsymbol{R}'')$ and rewrite this operator in the form

$$\mathsf{M}_*^R(\theta) = \sum_{\boldsymbol{R}'} m_\beta(\boldsymbol{R},\boldsymbol{R}') \left[\frac{y(\beta\boldsymbol{R}')}{h'(\theta)} - \frac{y(\beta\boldsymbol{R})}{h(\theta)}\right], \tag{8.5.10}$$

where

$$\begin{aligned}
m_\beta(\boldsymbol{R},\boldsymbol{R}') &= h(\theta)\theta(\beta\boldsymbol{R})m(\beta\boldsymbol{R},\beta\boldsymbol{R}')h'(\theta)q(\theta),\\
y(\beta\boldsymbol{R}) &= \frac{\theta(\beta\boldsymbol{R},t)}{\theta_R(\beta\boldsymbol{R})},\\
h(\theta) &= \frac{q(\theta)}{q(\theta_R)},\\
h'(\theta) &= \frac{q'(\theta)}{q'(\theta_R)},\\
q(\theta) &= [1 - y(\beta\boldsymbol{R})\theta_R(\beta\boldsymbol{R})]\, y(\beta-1\,\boldsymbol{R})\theta_R(\beta-1\,\boldsymbol{R}),\\
q'(\theta) &= [1 - y(\beta\boldsymbol{R}')\theta_R(\beta\boldsymbol{R}')]\, y(\beta-1\,\boldsymbol{R}')\theta_R(\beta-1\,\boldsymbol{R}'),\\
q(\theta_R) &= q(\theta)\big|_{y=1}.
\end{aligned}$$

It is an easy matter to check that

$$\begin{aligned}
m_\beta(\boldsymbol{R},\boldsymbol{R}') &= m_\beta(\boldsymbol{R}',\boldsymbol{R}),\\
L(\boldsymbol{R},\boldsymbol{R}') &\equiv \frac{y(\beta\boldsymbol{R})}{h(\theta)} - \frac{y(\beta\boldsymbol{R}')}{h'(\theta)} = -L(\boldsymbol{R}',\boldsymbol{R}),\\
\mathsf{M}_*^R(y)\big|_{y=1} &= 0.
\end{aligned} \tag{8.5.4}$$

Using standard algorithm (Dubrovskiy 1982) for the transformation of discrete operator (8.5.10) to diffusion form and taking into account (8.5.7) yield the following Fokker–Planck representation for M_*^R:

$$\mathsf{M}_*^R(\theta) = \sum_{ij} \partial_{R_i} \left[D_{ij}^R(\boldsymbol{R}|\theta) \left(\partial_{R_j} + \frac{1}{k_B T} \partial_{R_i} V(\boldsymbol{R}|\theta) \right) \right] \theta(\boldsymbol{\alpha}, t), \quad (8.5.12)$$

$$V(\boldsymbol{R}|\theta) = -k_B T \ln \theta_R \frac{q(\theta)}{q(\theta_R)}, \quad (8.5.13)$$

$$D_{ij}^R(\boldsymbol{R}|\theta) \approx q(\theta) \tilde{D}_{ij}^R(\boldsymbol{R}),$$

$$\tilde{D}_{ij}^R = \frac{1}{3} \sum_{\boldsymbol{R}''} R_i'' R_j'' m(\boldsymbol{R}, \boldsymbol{R} + \boldsymbol{R}''), \qquad R_i = (x, y), \quad (8.5.14)$$

In (8.5.12) the diffusion tensor D_{ij}^R is expressed through $q(\theta)$ and $m(\boldsymbol{R}, \boldsymbol{R}')$, the particle drift potential V – through $q(\theta)$ and the quasiequilibrium distribution function θ_R (that is defined up to an arbitrary function $\mu(\beta, t)$, see next chap.). For isotropic diffusion one may let

$$\tilde{D}_{ij}^R(\boldsymbol{R}) = \delta_{ij} \tilde{D}(\boldsymbol{R}). \quad (8.5.15)$$

The diffusion tensor D_{ij}^R that is proportional to $q(\theta)$ vanishes in the points where $\theta(\beta, \boldsymbol{R}, t) = 1$ and $\theta(\beta - 1, \boldsymbol{R}, t) = 0$ ($\beta \geq 3$). In these points $V \to \infty$, thus occupied cells and cells lying over unoccupied ones ($\beta \geq 3$) are automatically excluded from diffusion field that matches the basic postulates of the model. Operator $\mathsf{M}_*^R(\theta)$ vanishes when $\theta = \theta_R$ similar to its discrete analogue.

From the mathematical point of view equations (8.5.1) and (8.5.10) with definitions (8.5.3)–(8.5.5) form a system of nonlinear parabolic partial differential equations of the "Reaction + Diffusion" type, describing a wide range of physical phenomena that are of a great importance for thin films growth dynamics.

One can formally get a diffusion representation for the total migration operator $\mathsf{M}_*(\theta) = \mathsf{M}_*^R(\theta) + \mathsf{M}_*^\beta(\theta)$, coming to

$$\mathsf{M}_*(\theta) = \sum_{ij} \partial_{r_i} \left[D_{ij}(\boldsymbol{r}|\theta) \left(\partial_{r_j} + \frac{1}{k_B T} \partial_{r_i} V(\boldsymbol{r}|\theta) \right) \right] \theta(\boldsymbol{r}, t) \quad (8.5.16)$$

with

$$D_{ij}(\boldsymbol{r}|\theta) \approx q(\theta) \tilde{D}_{ij}(\boldsymbol{r});$$

$$V(\boldsymbol{r}|\theta) = -k_B T \ln \theta_R \frac{q(\theta)}{q(\theta_M)};$$

$$\tilde{D}_{ij} = \frac{1}{3} \sum_{\boldsymbol{r}''} r_i'' r_j'' m(\boldsymbol{r}, \boldsymbol{r} + \boldsymbol{r}''), \qquad \boldsymbol{r} \equiv \boldsymbol{\alpha}, \qquad r_i = (x, y, z);$$

$$q(\theta) = (1 - \theta(\boldsymbol{r}))\theta(\boldsymbol{r}_-), \qquad \boldsymbol{r}_- = \boldsymbol{r} - \boldsymbol{h}, \qquad \boldsymbol{h} = \boldsymbol{e}_z h_1.$$

In this case the kinetic equation of the thin film growth is transformed to the following form

$$\partial_t \theta(\boldsymbol{r}, t) = \mathsf{M}_*(\theta) + \mathsf{AD}_*(\theta, j). \quad (8.5.17)$$

9 Equilibrium and Quasiequilibrium Solutions to the LG Model

9.1 Equilibrium Distribution for Structureless Gas

To find the distribution function $\theta_c^0(\alpha)$ satisfying the conditions (8.2.24)–(8.2.27) one needs more detailed information on quantities r_u, q_u. This problem cannot be solved in a general form. Therefore let us consider first of all the case of structureless particles, when $\theta_c^0(\alpha)$ should be replaced with $\theta^0(\alpha)$ and the condition (8.2.27) is absent. One can readily prove that the quasiequilibrium solution meeting (8.2.24) can be presented in the following way

$$\theta^0(\alpha,t) = \theta_R(\alpha,\mu_R(\beta,t)) = \left[1+\zeta(\theta_R)\exp\left(\frac{\varepsilon(\alpha)-\mu_R(\beta,t)}{k_{\mathrm{B}}T}\right)\right]^{-1}, \quad (9.1.1)$$

$$\zeta(\theta_R) = \begin{cases} 1, & \beta = 1,2 \\ \theta_R^{-1}(\alpha_-), & \beta \geq 3 \end{cases},$$

$$\alpha_- = (\beta-1,\boldsymbol{R}), \qquad \alpha = (\beta,\boldsymbol{R}),$$

where $\mu_R(\beta,t)$ is an arbitrary function having a sense of the chemical potential of the β-th layer, and $T = T_s$ here and below unless otherwise specified.

The condition (8.2.25) leads to an analogous expression

$$\theta^0(\alpha,t) = \theta_\beta(\alpha,\mu_\beta(\boldsymbol{R},t)) = \left[1+\zeta(\theta_\beta)\exp\left(\frac{\varepsilon(\alpha)-\mu_\beta(\boldsymbol{R},t)}{k_{\mathrm{B}}T}\right)\right]^{-1}, \quad (9.1.2)$$

where an arbitrary function $\mu_\beta(\boldsymbol{R},t)$ has a sense of the chemical potential of the vertical column with a coordinate $\boldsymbol{R}$.

Evidently, when

$$\mu_R = \mu_\beta = \mu(t), \quad (9.1.3)$$

the distribution $\theta_R = \theta_\beta \equiv \theta_\mu(\alpha,\mu(t))$ vanishes the migration operator $\mathsf{M}_* = \mathsf{M}_*^R + \mathsf{M}_*^\beta$ and has the similar functional form.

From (8.2.26) an analogous dependence for the solution $\theta^0(\alpha) = \theta_a(\alpha,\mu_a)$ can be obtained, where μ_a is a constant that does not depend on α and is defined by microparameters σ, j, ν_1

$$\mu_a = k_{\mathrm{B}}T\ln\frac{\sigma j}{\nu_1}. \quad (9.1.4)$$

It is worth noting that the system of recurrent relations (9.1.1) or (9.1.2) has the form of the Fermi–Dirac distribution with quantities $\mu_R(\beta,t)$, $\mu_\beta(\beta,t)$, $\mu(t)$ and μ_a having the sense of a mean adsorption well depth. When $T\to 0$, $\theta_l(\alpha)\to 1$ provided $\varepsilon(\alpha) < \mu_l$ ($\mu_l = \mu_R$, μ_β, μ, μ_a) and $\theta_l(\alpha)\to 0$ provided $\varepsilon(\alpha) > \mu_l$ ($\varepsilon(\alpha) < 0$), i.e. for low temperatures the cells with the well depth exceeding μ_l are occupied, while more shallow wells turn out to be empty. With the growth of temperature this effect of selective occupation decreases.

The attractive lateral interactions between adatoms can be taken into account via introduction of the dependence of adsorption potential on the coverage

$$\varepsilon = \varepsilon(\alpha|\theta). \quad (9.1.5)$$

For example, using the mean field approximation (Kreuzer 1990) one should postulate

$$\varepsilon(\alpha|\theta) = \varepsilon_0(\alpha) - \phi_0\theta, \qquad \phi_0 > 0. \tag{9.1.6}$$

In case of complete equilibrium the distribution function $\theta_a(\alpha) = \theta_e(\alpha)$ within approximation (9.1.6) satisfies the following system of nonlinear equations:

$$b^0_\beta(1-\theta_a(\alpha))\theta_a(\alpha_-) = \theta_a(\alpha)\exp\left(-v_0\theta_a(\alpha)\right) \tag{9.1.7}$$

with

$$b^0_\beta = \frac{\sigma j}{\nu_1}\exp\left(-\frac{\varepsilon_0(\alpha)}{k_B T}\right); \quad v_0 = \frac{\phi_0}{k_B T}. \tag{9.1.8}$$

From the physical point of view equations (9.1.7) are nothing but a set of the coupled Fauler–Hugenheim isotherms for every layer

$$\frac{\theta_a(\alpha)}{1-\theta_a(\alpha)}\exp\left(-v_0\theta_a(\alpha)\right) = b^0_\beta\theta_a(\alpha_-), \tag{9.1.9}$$

with the form governed by the coverage of the underlying cells $\theta_a(\alpha_-)$ and by the values of parameters v_0 and b^0_β, where the latter can be rewritten as follows

$$b^0_\beta = \frac{j}{j_\beta} = \frac{P}{P_\beta}; \qquad P_\beta^{-1} = \frac{\sigma z}{\nu_1}\exp\left(-\frac{\varepsilon_0(\alpha)}{k_B T}\right); \tag{9.1.10}$$
$$j = zP; \qquad z = (2\pi m k_B T)^{-1/2}.$$

Note that the Fauler-Hugenheim isotherms take into account the first kind phase transitions in each layer.

It is useful to rewrite system (9.1.7) in the form

$$\theta_a(\alpha) = \frac{b^0_\beta\exp\left(v_0\theta_a(\alpha)\right)}{1+b^0_\beta\exp\left(v_0\theta_a(\alpha)\right)}, \qquad \beta = 1,2; \tag{9.1.11}$$

$$\theta_a(\alpha) = \frac{b^0_\beta\exp\left(v_0\theta_a(\alpha)\right)\theta_a(\alpha_-)}{1+b^0_\beta\exp\left(v_0\theta_a(\alpha)\right)\theta_a(\alpha_-)}, \qquad \beta \geq 3, \tag{9.1.12}$$

for the further analysis of main features of the equilibrium distributions defined by these equations.

9.2 Specific Features of Equilibrium Distributions

First of all it is essential to note that the quantities $j_\beta(\boldsymbol{R})$ and $P_\beta(\boldsymbol{R})$ determine the values of the flux density and pressure when the cell α is effectively occupied ($b^0_\beta = 1$). When $b^0_\beta \ll 1$ one has $\theta_a \ll 1$ (ideal adsorbate), while for $b^0_\beta \gg 1$ $\theta_a \approx 1$ and effects of nonideality become significant.

To simplify the analysis of (9.1.7) one may use an assumption that the first two layers are formed mainly by the substrate (model of chemisorption with precursor (Ceyer 1988)) and are characterized by different values of b^0_β, while for the upper layers ($\beta \geq 3$) b^0_β does not depend on β (BET-model):

$$b^0_1 \neq b^0_2 \neq b^0_3 = b^0_4 = \ldots = b^0_*; \qquad h_1 = h_2 = h_3 = \ldots. \tag{9.2.1}$$

Taking in (9.1.11), (9.1.12) $v_0 = 0$, $\theta_a(\alpha_-) = 1$ one comes to a set of the uncoupled Langmuir isotherms for ideal layers.

Under the conditions

$$v_0 = 0, \qquad b^0_\beta \theta_a(\alpha_-) \ll 1, \tag{9.2.2}$$

(9.1.12) leads to the conventional expression for the relative adsorbate volume $\Omega(\boldsymbol{R})$ (BET-model (Flood 1967))

$$\Omega(\boldsymbol{R}) = b^0_1 \left[(1 - b^0_*)(1 - b^0_* + b^0_1)\right]^{-1}. \tag{9.2.3}$$

For ideal adsorbate ($v_0 = 0$), provided condition (9.2.1) is valid, the system (9.1.7) has a unique solution that can be written in the form

$$\theta_a(\alpha) = b^0_\beta/(1 + b^0_\beta), \qquad \beta = 1, 2,$$
$$\theta_a(\alpha) = (b^0_* - 1) b^{0\,\beta-2}_* \theta_a(2, \boldsymbol{R}) \Big/ \left[b^0_*(b^{0\,\beta-2}_* - 1)\theta_a(2, \boldsymbol{R}) + b^0_* - 1\right], \quad \beta \geq 3 \tag{9.2.4}$$

Solution (9.2.4) is valid for arbitrary b^0_* values. In particular, when $b^0_* < 1$ the following simple approximation for $\theta_a(\alpha)$ with $\beta \geq 3$ is useful

$$\theta_a(\alpha) = b^{0\,\beta-2}_* \tilde{\theta}_a(2, \boldsymbol{R}), \qquad \tilde{\theta}_a(2, \boldsymbol{R}) = \theta_a(2, \boldsymbol{R}) \frac{b^0_* - 1}{b^0_* - 1 - b^0_* \theta_a(2, \boldsymbol{R})}. \tag{9.2.5}$$

When $b^0_* \ll 1$ $\tilde{\theta}_a(2, \boldsymbol{R}) = \theta_a(2, \boldsymbol{R})$ and (9.2.5) leads to the BET isotherm (9.2.3).

For nonideal adsorbate ($v_0 \neq 0$) different situations are possible. Under the condition

$$v_0 < \min\left(1 + (e b^0_\beta)^{-1}, 4\right) \tag{9.2.6}$$

system (9.1.7) has the unique solution that is close to zero (weakly nonideal adsorbate) and turns into (9.2.4) provided $v_0 = 0$ and $b^0_* < 1$. When the conditions

$$b^0_\beta > (3e)^{-1}, \qquad 1 + (e b^0_\beta) < v_0 < 4 \tag{9.2.7}$$

are met, one gets the unique solution that is close to the unity (filled adsorbate). If, at last,

$$b^0_\beta < (3e)^{-1}, \qquad 4 < v_0 < 1 + (e b^0_\beta), \tag{9.2.8}$$

there are three solutions. The first is close to zero, the second is close to unity, while the third one takes intermediate value. Such behavior of an isotherm is characteristic for the phase transition region.

From the above said it follows that the quantities b^0_β control the degree of the filling of α-th cell, while the parameter v_0 describes the "distance" to the phase transition point. For $b^0_\beta \to 0$, $v_0 \to \infty$ or $b^0_\beta \to \infty$, $v_0 \to -\infty$ equations (9.1.7) are met identically. Physically it means, that when the coverage parameter b^0_β decreases, the adsorbate growth is stimulated by the increase of the attractive lateral interaction between adatoms, and when b^0_β increases, the multilayer adsorbate can be formed even for the repulsive adatom–adatom interaction.

One more conclusion from the above analysis is that the conditions (9.2.7) and (9.2.8) can be considered as a microscopic criteria of the layer-by-layer and island growth mechanisms.

Let us discuss now in more detail the equilibrium solution for the case of multicomponent and chemically reacting gases. By analogy with (9.1.6) the mean field approximation may be introduced with the help of the relations

$$\varepsilon_c(\alpha|\theta_c) = \varepsilon_{c0}(\alpha) - \sum_{c'} \phi_{cc'}\theta_{c'}(\alpha), \qquad \phi_{cc'} > 0. \tag{9.2.9}$$

As it can be shown using (8.2.24)–(8.2.26), under equilibrium conditions these relations lead to the following system of recurrent relations analogous to (9.1.11) and (9.1.12)

$$\theta_{c0}(\alpha) = \frac{b_c^0 \prod_{c'} \exp\left(v_{cc'}\theta_{c'0}(\alpha)\right)}{\left[1 + \sum_{c'} b_{c'}^0(\beta) \prod_{c'} \exp\left(v_{cc'}\theta_{cc'}(\alpha)\right)\right]}, \quad \beta = 1, 2; \tag{9.2.10}$$

$$\theta_{c0}(\alpha) = \frac{b_c^0 \prod_{c'} \exp\left(v_{cc'}\theta_{c'0}(\alpha)\right) \sum_{c''} \theta_{c''}(\alpha_-)}{\left[1 + \sum_{c'} b_{c'}^0(\beta) \prod_{c'} \exp\left(v_{cc'}\theta_{cc'}(\alpha)\right) \sum_{c''} \theta_{cc''}(\alpha_-)\right]}, \quad \beta \geq 3. \tag{9.2.11}$$

Here $v_{cc'} = \phi_{cc'}/k_B T$, $b_c^0 = \exp\left((\mu_c - \varepsilon_{co}(\beta))/k_B T\right)$,

$$\mu_c = k_B T \ln((j_c\sigma_c)/\nu_{1c}), \tag{9.2.12}$$

with parameter μ_c playing the role of the chemical potential of c-th component in equilibrium.

For ideal adsorbate ($v_{cc'} = 0$) the formulae (9.2.10) and (9.2.11) may be significantly simplified. Particularly, under an approximation

$$b_c^0(1) \neq b_c^0(2) \neq b_c^0(3) = b_c^0(4) = \ldots b_{c*}^0 \tag{9.2.13}$$

one can obtain an analogous to (9.2.4) solution in explicit form

$$\theta_{c0}(\alpha) = b_c^0(\beta)/[1 + b_c^0(\beta)], \qquad \beta = 1, 2; \tag{9.2.14}$$

$$\theta_{c0}(\alpha) = b_{c*}^0 L(b_*^0, \theta(2, \boldsymbol{R})), \qquad \beta \geq 3, \tag{9.2.15}$$

with $b_*^0 = \sum_c b_{c*}^0$, $\theta(2, \boldsymbol{R}) = \sum_c \theta_{c0}(2, \boldsymbol{R})$ and

$$L(x, y) = x^{\beta-3}(x-1)y \Big/ \left\{x - 1 + [x^{\beta-2}(x-1) + x(x^{\beta-3} - 1)]y\right\}. \tag{9.2.16}$$

Let us now include into consideration surface chemical reactions. The solution (9.2.15) with an eye to (8.2.27) leads to a set of relations

$$\frac{r_u}{q_u} \prod_c \exp\left(\frac{\mu_c}{k_B T}\gamma_{cu}\right) L^{\gamma_{cu}} = C_u, \quad u = 1, 2 \ldots u_0, \quad \gamma_{cu} = 0 \text{ or } 1 \tag{9.2.17}$$

with C_u being a constant specific for each of the reactions. The further analysis of (9.2.17) requires further details concerning surface reactions.

As an example of the above discussed results application, let us consider the evaluation of the quasiequilibrium solution for the case of fast diffusion of structureless particles based on singularly perturbed equation (8.3.7).

Expanding the solution in powers of small parameter ε gives

$$\theta(\alpha,t) = \theta_M(\alpha,t) + \varepsilon\theta_1(\alpha,t) + \varepsilon^2\theta_2(\alpha,t) + \dots. \tag{9.2.18}$$

Inserting this expansion into (8.3.7) yields the following system for determining θ_M, θ_1, θ_2, etc.:

$$\begin{aligned} \mathsf{M}(\theta_M) &= 0; \\ \partial_t\theta_M - \mathsf{AD}(\theta_M, j) &= \mathsf{M}_1(\theta_1); \\ \partial_t\theta_1 - \mathsf{AD}_1(\theta_1, j) &= \mathsf{M}_2(\theta_2), \qquad \dots \end{aligned} \tag{9.2.19}$$

where M_k, AD_k ($k = 1, 2, \dots$) are linearized with respect to θ_k operators M, AD. From the first equation in (9.2.19) it follows that θ_μ meets (9.1.2) where $\mu(t)$ is an arbitrary function of time. From the condition of the existence of a solution to the second equation in (9.2.19) one can obtain the following evolution equation for $\mu(t)$

$$\frac{d\mu}{dt} = P(\mu)\left[\exp\left(\frac{\mu_a - \mu(t)}{k_B T}\right) - 1\right], \qquad P(\mu) > 0, \tag{9.2.20}$$

where we do not specify particular approximations for the function $P(\mu)$.

Equation (9.2.20) describes the relaxation of the parameter $\mu(t)$ to the equilibrium value μ_a. A quasiequilibrium solution in this case is build on the base of the results of previous Secs. and of (9.2.20). For the other two cases quasiequilibrium solutions can be treated in a similar way.

9.3 Evaluation of Characteristics of Growing Film

The LG model allows one to evaluate both microscopic relief and average macroparameters of a growing film on the base of definitions (6.1.10). As an example let us consider the problem of macrorelief $h(\boldsymbol{R}, t)$ calculation for the case of structureless particles. Starting from definitions (6.1.10) and kinetic equation (8.5.1) one can derive evolution equations for the macroprofile $h(\boldsymbol{R}, t)$ and its mean variance $\Delta(\boldsymbol{R}, t)$:

$$\partial_t h(\boldsymbol{R}, t) = -h(\boldsymbol{R}, t)\partial_t \ln Q(\boldsymbol{R}, t) + \mathsf{M}_h(h) + \mathsf{AD}_h(h, j), \tag{9.3.1}$$

$$\partial_t \Delta(\boldsymbol{R}, t) = -\Delta(\boldsymbol{R}, t)\partial_t \ln Q(\boldsymbol{R}, t) + \mathsf{M}_\Delta(\Delta) + \mathsf{AD}_\Delta(\Delta, j), \tag{9.3.2}$$

where the following notation has been introduced

$$\mathsf{M}_h(h) = Q^{-1}\sum_{\beta\geq 0}\beta\mathsf{M}_*(\theta), \quad \mathsf{AD}_h(h, j) = Q^{-1}\sum_{\beta\geq 0}\beta\mathsf{AD}_*(\theta, j); \tag{9.3.3}$$

$$\mathsf{M}_\Delta(\Delta) = Q^{-1}\sum_{\beta\geq 0}(\beta - h)^2\mathsf{M}_*(\theta), \quad \mathsf{AD}_\Delta(\Delta) = Q^{-1}\sum_{\beta\geq 0}(\beta - h)^2\mathsf{AD}_*(\theta). \tag{9.3.4}$$

Mathematically quantities $Q(\boldsymbol{R}, t)$, $h(\boldsymbol{R}, t)$, $\Delta(\boldsymbol{R}, t)$, $\bar{h}(t)$, $\bar{\Delta}(t)$, etc. are nothing but the momenta of the distribution function $\theta(\alpha, t)$, therefore (9.3.1)–(9.3.2) can be completed in the framework of some approximate solution to the basic kinetic equation (8.5.1) (obtained within asymptotic or linearization methods, finite–dimensional approximation of kinetic operators and so on). The appropriate boundary and initial conditions are to be formulated on the base of experimental data or theoretical considerations (the problem of initial and boundary layers). On the other hand, the quantities h, Δ can be calculated directly from (6.1.10) provided some approximate form of the distribution function $\theta(\alpha, t)$ is known. Let us present the examples of analytical calculations of h under some simple approximations.

In equilibrium (or quasiequilibrium) using approximation (9.2.5) ($v_0 = 0$, $b_*^0 < 1$) one can obtain from (6.1.10) the following expression for the equilibrium microprofile $h_e(\boldsymbol{R})$

$$h_e = \left(f_{1a} + \tilde{\theta}_a(2) \frac{b_*^0(3 - 2b_*^0)}{(1 - b_*^0)^2} \right) \Big/ \left(f_{2a} + \tilde{\theta}_a(2) \frac{b_*^0}{1 - b_*^0} \right). \qquad (9.3.5)$$

$$f_{1a} = \theta_a(1) - 2\theta_a(2); \qquad f_{2a} = 1 + \theta_a(1)\theta_a(2).$$

This solution describes the following physical effects. When $b_*^0 \ll 1$ (a rarefied multilayer adsorbate) $\tilde{\theta}_a(2) \approx \theta_a(2)$ and h_e is defined entirely by the coverages of the first two layers. If $b_*^0 \to 1$

$$h_e \sim \frac{1}{1 - b_*^0} \to \infty, \qquad (9.3.6)$$

i.e. the film height grows unlimitedly and the film quality does not depend on the substrate.

For $b_*^0 > 1$ and $v_0 = 0$ one has $\theta_\beta \approx (b_*^0 - 1)/b_*^0 = \theta_*$ and $h_e \to \infty$. Thus, $b_*^0 = 1$ is a singular point for the dynamics of film growth.

For nonideal adsorbate ($v_0 \neq 0$) analytical formulae for h_e cannot be obtained. Nevertheless, simple numerical investigation shows that if the system (9.1.7) has a unique solution, the film has either finite height or grows unlimitedly depending on the value of the parameter b_*^0. If (9.1.7) has several solutions, in the limit $b_*^0 \to 1$ one encounters a complicated situation (bifurcation) that needs detailed physical and mathematical investigation.

One can readily obtain closed kinetic equations for $h(\boldsymbol{R}, t)$ on the base of (9.3.1) and (8.5.1) provided lateral interactions and nonlinearity in the operators $\mathsf{M}_*(\theta)$, $\mathsf{AD}_*(j, \theta)$ are neglected. These equations have different forms depending on the values of the parameter b_*^0. In particular, in case $b_*^0 < 1$ we have

$$\partial_t h = -hg_1 + A_* + g_2, \qquad \mathsf{M}_h(h) \approx 0, \partial_t Q = B_*(b_*^0 - 1) + A_*\theta_2 + \partial_t f_2 + f_2(B_* - A_*) \qquad (9.3.7)$$

with

$$g_1 = Q^{-1}\left[\partial_t(\theta_1\theta_2) + B_*(1 - b_*^0)(1 + \theta_1\theta_2 + A_*\theta_2\right], \qquad (9.3.8)$$

$$g_2 = Q^{-1}\left[A_*(\theta_2 - \theta_1 - \theta_1\theta_2 - 1) + B_*(\theta_1 + 2\theta_2) + \partial_t(\theta_1 + 2\theta_2)\right], \qquad (9.3.9)$$

$$\theta_\beta = \theta_{\beta e} + (\theta_\beta^0 - \theta_{\beta e} \exp\left[-(A_\beta + B_\beta)t\right]),$$

$$A_\beta = j\sigma a(\beta),$$

$$B_\beta = \nu_1 a(\beta) \exp\left(\frac{\varepsilon(\beta)}{k_B T}\right), \qquad \beta = 1, 2.$$

In (9.3.7)–(9.3.9) the following notation is used

$$A_* = j\sigma_* a_*, \quad B_* = \nu_{1*} a_* \exp\left(\frac{\varepsilon_*}{k_B T}\right), \quad b_*^0 = A_*/B_*,$$

$$\theta_\beta \equiv \theta(\beta, \boldsymbol{R}, t), \quad \theta_\beta^0 \equiv \theta(\beta, \boldsymbol{R}, 0), \quad \theta_{\beta e} = b_\beta^0/(1 + b_\beta^0),$$

$$b_\beta^0 = A_\beta/B_\beta \qquad (9.3.10).$$

These formulae allow one to prove that for $b_*^0 < 1$ $h(t) \to h_a$ when $t \to \infty$, where h_a is defined by (9.3.5), while for $b_*^0 > 1$ in the limit $t \to \infty$ it is possible to prove that

$$h(t) \approx ct, \tag{9.3.11}$$

i.e. the unlimited film growth takes place.

We do not present here more sophisticated equations for $h(\boldsymbol{R}, t)$, neither do we consider the question of $h(\boldsymbol{R}, t)$ dependence on initial conditions (initial geometric inhomogeneity of the substrate). These important in applications questions are more complicated and need a special analysis.

As it has been shown, the model qualitatively describes a number of features of growth dynamics for ideal and nonideal adsorbate. Possible generalizations of this model can be based on more precise adsorption isotherms of multilayer adsorbate (calculated in other than the mean field approximation) and including more realistic diffusion and adsorption coefficients dependence on the coverage.

9.4 Statistical Evaluation of Equilibrium Parameters of Adsorbed Films

In this section we present an example of a more rigorous evaluation of equilibrium state parameters of adsorbed film by statistical mechanics methods (Dubrovskiy V. 1990). The aim is, at first, to show how starting from the potential of adatom – surface interaction one can substantiate different models. Secondly, this example shows the consequences of including lateral interactions between adatoms for adsorption isotherms.

Let us model the interaction potential by the expression

$$V(z, \boldsymbol{R}) = V(z_m, \boldsymbol{R}_m) + \frac{1}{2}c_z(z - z_m)^2 + V_R(\boldsymbol{R}), \tag{9.4.1}$$

that is equivalent to the quadratic approximation in z of the potential at the minimum point $(z_m, \boldsymbol{R}_m)$. The constant c_z is supposed to be weakly dependent on $\boldsymbol{R}$, therefore the two-dimensional adsorbed phase is flat and the function $V_R(\boldsymbol{R}) = V(z_m, \boldsymbol{R}) - V(z_m, \boldsymbol{R}_m) > 0$ describes the potential barrier for lateral diffusion (see Fig. 4). When considering classical configuration integral Q_N for canonical ensemble of N particles, one can apply the Laplacf method over z coordinate for its evaluation if temperatures are low enough to provide $|V(z_m, \boldsymbol{R}_m)|/k_B T$ and $c_z z_m^2/2k_B T$ being large. This will result in

$$Q_N = \left[z_f \exp\left(-\frac{\varepsilon}{k_B T} \right) \right]^N Q_N^{(2D)}(T, S), \tag{9.4.2}$$

$$Q_N^{(2D)} = \frac{1}{N!} \int_S d\boldsymbol{R}_1 \ldots d\boldsymbol{R}_N \times \exp\left[-\frac{1}{k_B T} \left(\sum_{1 \le i < j \le N} V(R_{ij} + \sum_{i=1}^{N} V_R(\boldsymbol{R}_i) \right) \right], \tag{9.4.3}$$

where $-\varepsilon = -V(z_m, \boldsymbol{R}_m) > 0$, $z_f(T) = (2\pi k_B T/c_z)^{1/2}$ is the amplitude of adatom vibration in normal direction, S – substrate surface area, $Q_N^{(2D)}$ is configuration integral of two-dimensional gas. Adsorbed monolayers can be free ($-\varepsilon \geq k_{\rm B}T \gg \varepsilon_D$), pseudolocalized ($-\varepsilon \geq k_{\rm B}T \approx \varepsilon_D$) or strongly localized ($k_{\rm B}T \ll \varepsilon_D \approx \varepsilon$). The latter case of localized adsorption is typical for thin film growth on crystalline substrates.

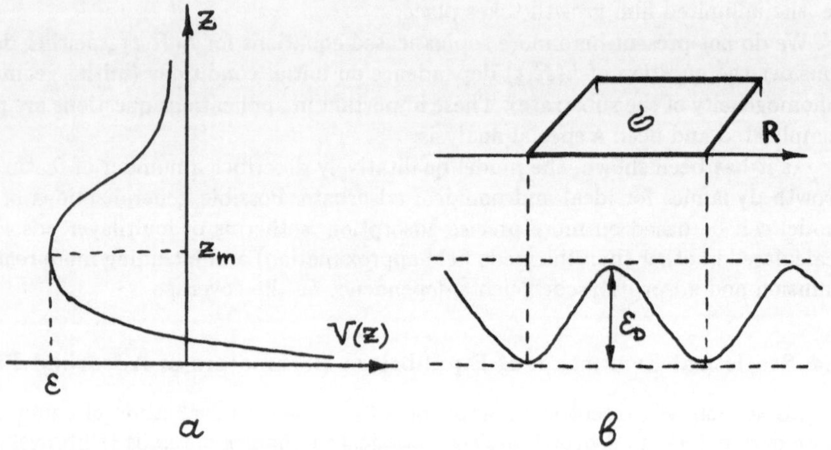

Fig. 4. Typical shape of the collective potential of adatom–surface interaction. a) the shape of potential along z-axis; b) schematic shape of potential in plane $z = z_m$; ε – adsorption potential, $\varepsilon_{\rm D}$ – diffusion barrier, σ – elementary adsorption cell area.

Let us consider a two-dimensional lattice of adsorption centers of one type localized at the nodes with coordinates $\boldsymbol{a}_i$. Then one can write

$$\boldsymbol{R}_i = \boldsymbol{a}_i + \boldsymbol{R}'_i, \ \ R_{ij} = (a_{ij}^2 + 2\boldsymbol{a}_{ij}\boldsymbol{R}'_{ij} + {\boldsymbol{R}'_{ij}}^2)^{\frac{1}{2}}, \tag{9.4.4}$$

where vector $\boldsymbol{R}'_i$ lies inside the elementary cell. Thus integral over the surface area can be presented as the sum over surface cells and integration over the unit cell of area σ

$$\int_S \mathrm{d}\boldsymbol{R}_i = \sum_{\boldsymbol{a}_i} \int_\sigma \mathrm{d}\boldsymbol{R}'_i \tag{9.4.5}$$

We shall further define a quantity

$$\sigma_f = \int_\sigma \mathrm{d}\boldsymbol{R}' \exp\left(-\frac{V_R(\boldsymbol{R}')}{k_{\rm B}T}\right), \tag{9.4.6}$$

that has the sense of the mean area available to a particle through lateral motion ($\sigma_f/\sigma < 1$). It is obvious, that $p(\boldsymbol{R}) = \sigma_f^{-1}\exp(-V_R(\boldsymbol{R})/k_{\rm B}T)$ is the probability density of finding adatom in the vicinity of $\boldsymbol{R}$ within the limits of one cell. In terms of σ_f the configuration integral $Q_N^{(2D)}$ can be presented in the form

$$Q_N^{(2D)}(T, N_c, \sigma) = \frac{1}{N!} \sum_{\boldsymbol{a}_1 \ldots \boldsymbol{a}_{N_c}} \int_\sigma \mathrm{d}\boldsymbol{R}'_1 \ldots \mathrm{d}\boldsymbol{R}'_N p(\boldsymbol{R}'_1) \ldots p(\boldsymbol{R}'_N)$$

$$\times \exp\left(-\frac{1}{k_{\rm B}T} \sum_{1 \leq i < j \leq N} V(R_{ij})\right) \sigma_f^N, \tag{9.4.7}$$

where N_c is a number of adsorption cells.

Provided the interaction potential near points (a_k, z_m) depends weakly both on $\boldsymbol{R}'$ and z $(\sigma_f/\sigma \ll 1)$, the integrals over $\boldsymbol{R}'$ may be evaluated by the Laplace method giving

$$Q_N = \left[v_f \exp\left(-\frac{\varepsilon}{k_B T}\right)\right]^N \frac{1}{N!} \sum_{a_1 \ldots a_{N_c}} \exp\left(-\frac{1}{k_B T} \sum_{1 \le i < j \le N} V(a_{ij})\right), \quad (9.4.8)$$

$$a_{ij} = |\boldsymbol{a}_i - \boldsymbol{a}_j|, \quad v_f = z_f \sigma_f = \left(\frac{2\pi k_B T}{c_z}\right)^{\frac{1}{2}} \int_\sigma \mathrm{d}\boldsymbol{R}' \exp\left(-\frac{V_R(\boldsymbol{R}')}{k_B T}\right), \quad (9.4.9)$$

with v_f being the volume available to a particle moving near the adsorption site.

Expression (9.4.8) presents the configuration integral of two–dimensional lattice gas. Introducing cell occupation numbers $n_i = (0, 1)$ $(N = \sum_{i=1}^{N_c} n_i)$ gives for the lattice gas partition function

$$Z_N = \left[\frac{v_f}{\lambda^3} \exp\left(-\frac{\varepsilon}{k_B T}\right)\right]^N \sum_{n_1 \ldots n_{N_c} = N} \exp\left[\frac{1}{k_B T} \sum_{1 \le i < j \le N} V(a_{ij}) n_i n_j\right] \quad (9.4.10)$$

with $\lambda(T) = h/(2\pi \mu k_B T)^{\frac{1}{2}}$.

Including repulsive part V^r of the interaction potential via condition

$$V^r(a_{ij}) = \begin{cases} 0, & i \neq j \\ \infty, & i = j \end{cases} \quad (9.4.11)$$

and attractive part V^a via the mean field approximation

$$V^a = -\phi_0 \equiv \sum_{i \neq 0} V(a_{0i}), \quad (9.4.12)$$

yields the following self-consistent expression for the partition function

$$Z_N = \frac{N_c!}{N!(N_c - N)!} Q_0^N \left(1 + \frac{v_0}{2} \frac{N^2}{N_c}\right), \quad (9.4.13)$$

$$v_0 = \frac{\phi_0}{k_B T}, \quad Q_0 = v_f \lambda^{-3} \exp\left(-\frac{\varepsilon}{k_B T}\right), \quad \phi_0 > 0. \quad (9.4.14)$$

For the chemical potential of two-dimensional gas one then obtains an expression

$$\mu_a = -k_B T \frac{\partial \ln Z_N}{\partial N} = -k_B T \left[\ln\left(\frac{1-\theta}{\theta}\right) + \ln Q_0 + v_0 \theta\right], \quad (9.4.15)$$

with $\theta = N/N_c$ being the substrate coverage.

Using the ideal gas chemical potential $M_g = k_B T \ln(P/k_B T) + \ln \Lambda^3$ and the condition of the equilibrium between two-dimensional adsorbate and gaseous phase $(\mu_d = \mu_a)$ leads to the Fauler–Hugenheim isotherm (Dash 1975; Jaycock and Parfitt 1981)

$$bP = \frac{\theta}{1-\theta} \exp(-v_0 \theta), \quad b = \frac{v_f}{k_B T} \exp\left(-\frac{\varepsilon}{k_B T}\right). \quad (9.4.16)$$

Provided $\phi_0 = 0$, (9.4.16) turns into the Langmuir isotherm.

For the most part of thin film deposition technologies the equilibrium conditions are not met ($T \neq T_s$, $\theta = \theta(t)$, $S = S(t)$). Nevertheless, the study of the interphase equilibrium in two-dimensional adsorbate is necessary for the description of the phase transition kinetics. The investigation of (9.4.15) results in a qualitative picture presented in Fig. 5. The critical temperature T_c is

$$T_c = \frac{\phi_0}{4k_B} \quad (v_0 = 4). \tag{9.4.17}$$

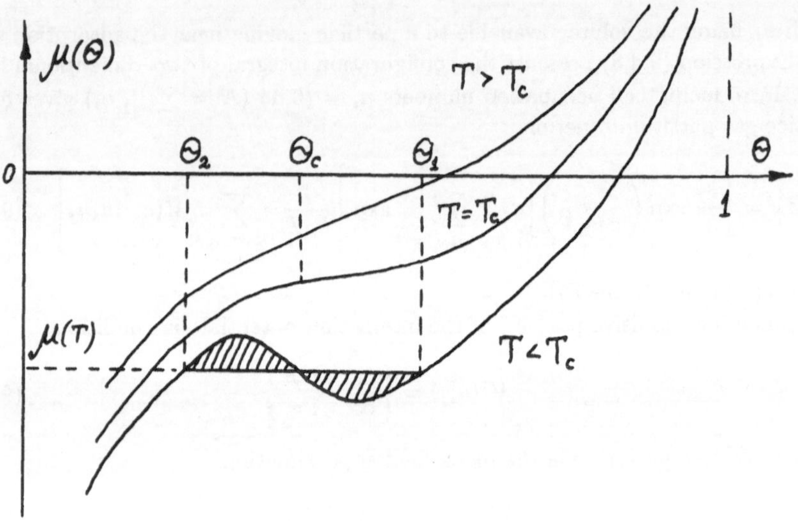

Fig. 5. Chemical potential μ_a dependence on surface coverage θ: μ_e is the equilibrium value of the chemical potential; $\theta_{2,1}$ – the equilibrium values of surface coverage by phase 2 (vapor) and 1 (condensate); θ_c is the critical coverage.

The condition of the thermodynamical stability of adsorbate ($\mathrm{d}M(\theta)/\mathrm{d}\theta \geq 0$) requires the replacement of the wavy part of the curve in Fig. 5 for $T < T_c$ by a horizontal line $\mu_a = \mu_e(T)$ (the Maxwell procedure) such that

$$\mu_e(T) = \mu_a(\theta_2^0) = \mu_a(\theta_1^0) = (\theta_1^0 - \theta_2^0)^{-1} \int_{\theta_2^0}^{\theta_1^0} \mathrm{d}\theta' \mu_a(\theta'). \tag{9.4.18}$$

It is easy to show that the Maxwell's rule is equivalent to the condition of the equality of the surface energies of the phases 2 and 1 (vapor and condensate)

$$q(\theta_2) = q(\theta_1). \tag{9.4.19}$$

Three equilibrium conditions $\mu_e = \mu_a(\theta_2) = \mu_a(\theta_1)$; $q(\theta_2) = q(\theta_1)$ allow one to obtain three unknown quantities μ_e, θ_2, θ_1. For example, from (9.4.15) for low temperatures of substrate ($\theta_2 \ll 1$) one can get the following expression for the coverage of saturated vapor of adsorbed particles

$$\theta_2 = \exp\left(\frac{\phi_0}{2k_B T}\right), \tag{9.4.20}$$

that represents a natural analogue of the Clapeyron - Clausius formula. Thus, for $T < T_c$ and fixed potential μ_e the co-existence of the three phases: gas with pressure $P = \Lambda^{-3} k_B T \exp(\mu_e / k_B T)$, phase 1 and phase 2 is possible. For $T > T_c$ only one of the monolayer phases can be in equilibrium with the gas.

10 Thermodynamical Description of Adsorbed Layer Dynamics

10.1 Description of Adlayer Dynamics by Methods of Statistical Thermodynamics

The equations of isothermal kinetics (8.3.7)-(8.3.9) for the one-particle distribution function can include lateral interactions both via the dependence of the transition probabilities on the coverage (the mean-field approximation) and via direct correlations between elementary processes in different cells. The strict consideration of the problem of lateral interaction may be given by methods of statistical thermodynamics.

Let us consider the main ideas of these methods on an example of a structureless lattice gas of monolayer adsorbate. The distribution of adatoms over discrete cells can be described by a many-particle distribution function $F(n, t)$ $(n = (n_1, n_2 \ldots n_k)$, $k = N_c)$, where n_j is the occupation number of a cell number j $(n_j = 0, 1$ for single-place cells), N_c - the total number of cells that is considered to be a constant (we neglect here the processes of surface reconstruction). The distribution function $F(n, t)$ is normalized by the condition

$$\sum_n F(n, t) = 1, \tag{10.1.1}$$

where the summation runs over all possible configurations of the filled cells Thus, the distribution function $F(n, t)$ has the meaning of configurations density distribution (many-particle distribution function in the occupation number space). It will be assumed that the sequence of different configurations is a Markov random process initiated by elementary acts of desorption, adsorption, and diffusion. It will be assumed also that the distribution function obeys a standard master equation

$$\partial_t F(n, t) = \mathsf{I}_*(F) \equiv \sum_{n'} \left[W(n', n) F(n', t) - W(n, n') F(n, t)\right], \tag{10.1.2}$$

where $W(n', n)$ is the probability of a transition from configuration n' to n per unit time. Under the assumption that there is an energy exchange of the adsorbate with the substrate and the gas phase and there is a particle exchange between the adsorbate and the gas phase the equilibrium state of the adsorbate can be described by the following equilibrium distribution function $F_0(n)$

$$F_0(n) = Z_F^{-1} \exp\left(-\frac{H_0(n)-\mu N}{k_B T}\right), \tag{10.1.3}$$

$$N_a = \sum_{j=1}^{N_c} n_j, \quad H_0(n) = \sum_{j=1}^{N_c} \varepsilon_j n_j + \frac{1}{2}\sum_{j,k=1}^{N_c} \omega_{2jk} n_j n_k, \tag{10.1.4}$$

$$Z_F = \sum_n \exp\left(-\frac{H_0(n)-\mu N}{k_B T}\right). \tag{10.1.5}$$

Here $H_0(n)$ is the thermodynamical Hamiltonian of the adsorbate, ε_j is the the adsorption potential of the j-th cell, ω_{2jk} is the energy of the pair interaction between the j-th and k-th particles, μ is the chemical potential of the adsorbate, and Z_F is the adsorbate partition function. Introducing the thermodynamical expression for adatom energy in adsorption cell

$$\varepsilon_j - \mu \equiv \omega_{1j}, \tag{10.1.6}$$

yields for $F_0(n)$ and Z_F

$$F_0(n) = Z_F^{-1} \exp\left(-\frac{H(n)}{k_B T}\right), \quad Z_F = \sum_n \exp\left(-\frac{H(n)}{k_B T}\right), \tag{10.1.7}$$

$$H(n) = \sum_{j=1}^{N_c} \omega_{1j} n_j + \frac{1}{2}\sum_{j,k=1}^{N_c} \omega_{2jk} n_j n_k. \tag{10.1.8}$$

In order that the expression (10.1.7) be a unique solution to (10.1.2) for $t \to \infty$, the probabilities $W(n',n)$, $W(n,n')$ are to obey the microscopic detailed balance principle

$$F_0(n')W(n',n) = F_0(n)W(n,n'). \tag{10.1.9}$$

The following expression can be used for ω_{1j}

$$\omega_{1j} = \varepsilon_j - k_B T \ln Z_{3j} - k_B T \ln \frac{\lambda^3 P}{k_B T}. \tag{10.1.10}$$

In case of a homogeneous adsorbate one has

$$\omega_{1j} = \omega_1, \quad \omega_{2jk} = \omega_2. \tag{10.1.11}$$

The following quantities can be defined in terms of the many-particle distribution function $F(n,t)$: the mean occupancy of the j-th cell

$$\theta_j(t) \equiv < n_j > \equiv \sum_n n_j F(n,t), \tag{10.1.12}$$

the mean layer coverage

$$\theta(t) = N_c^{-1} \sum_j \theta_j(t), \tag{10.1.13}$$

the two-, three-, four-, etc. particle distribution functions

$$\theta_{jk} \equiv < n_j n_k >, \quad \theta_{jkl} \equiv < n_j n_k n_l >, \quad \theta_{jklm} \equiv < n_j n_k n_l n_m >, \tag{10.1.14}$$

the two-, three-, etc. particle correlation functions

$$\Psi_{jk} \equiv \theta_{jk} - \theta_j\theta_k, \quad \Psi_{jkl} \equiv \theta_{jkl} - \theta_{jk}\theta_l - \theta_{jl}\theta_k - \theta_{kl}\theta_j - \theta_j\theta_k\theta_l. \qquad (10.1.15)$$

The main question in this approach is the evaluation of the transition probabilities $W(n',n)$. For this purpose let us assume, that at the same time only single-particle acts of adsorption, desorption and diffusion can occur. Under this assumption the transition probabilities can be expanded following the ideas by Asada (Asada 1990). The elementary act of adsorption into the j-th cell with an eye to lateral interactions contributes to $W(n',n)$ as follows

$$W_{aj}(n',n) = W_j(1-n'_j)[1 + a_1 \sum_k n_{j+k} + a_2 \sum_{k,k'} n_{j+k}n_{j+k'}$$

$$+a_3 \sum_{k,k',k''} n_{j+k}n_{j+k'}n_{j+k''} + \ldots]\delta(n'_j, 1-n_j)\prod_{l\neq j}\delta(n'_l, n_l). \qquad (10.1.16)$$

The desorption from the j-th cell process contribution is

$$W_{dj}(n',n) = W_j d_0 n_j[1 + d_1 \sum_k n_{j+k} + d_2 \sum_{k,k'} n_{j+k}n_{j+k'}$$

$$+d_3 \sum_{k,k',k''} n_{j+k}n_{j+k'}n_{j+k''} + \ldots]\delta(n'_j, 1-n_j)\prod_{l\neq j}\delta(n'_l, n_l). \qquad (10.1.17)$$

A diffusion jump from the j-th cell to the k-th one makes the following contribution

$$W_{Dj,j+k}(n',n) = W_{j,j+k}n'_j(1-n'_{j+k})[1 + D'_1 \sum_{l\neq k} n_{j+l}$$

$$+D_1 \sum_{l\neq -k} n_{j+k+l} + D'_2 \sum_{l,l'\neq k} n_{j+l}n_{j+l'} + D_2 \sum_{l,l'\neq -k} n_{j+k+l}n_{j+k+l'} + \ldots]$$

$$\times\delta(n'_j, 1-n_j)\delta(n'_{j+k}, 1-n_{j+k})\prod_{l\neq j}\delta(n'_l, n_l)\prod_{l\neq j+k}\delta(n'_l, n_l). \qquad (10.1.18)$$

In the expansions (10.1.16)–(10.1.18) $\delta(n,m)$ is the Kroneker symbol; W_j, $W_j d_0$, $W_{j,j+k}$ are the one-particle probabilities of adsorption into the j-th cell, desorption from the j-th cell, and diffusion jump from the j-th to the $j+k$-th cell, respectively. Coefficients a_k, d_k, D'_k, D_k characterize the influence of lateral interactions with one, two, etc. adatoms on adsorption, desorption, and diffusion processes. The advantage of the representations (10.1.16)–(10.1.18) is that the parameters can be rigorously evaluated either from dynamical models ($W_j, W_{j,j+k}$) or from statistical thermodynamics (d_0, a_k, d_k, D'_k, D_k). Particularly, for the evaluation of d_0, a_k, d_k, D'_k, D_k one can use the equilibrium distribution function F_0 (such method has been widely used in statistical physics (Glauber 1963) and int the theory of temperature Green's functions (Kadanoff and Baym 1962)), the detailed balance conditions (10.1.9), the connection of this approach to Ising model (the latter connection becomes obvious after introducing spin variables $\sigma_j = 2n_j - 1$). Let us present some examples of the statistical parameters evaluation.

Using the detailed balance condition (10.1.9) together with (10.1.16) and (10.1.17) gives (Asada 1990)

$$d_0 = \exp\left(\frac{\omega_1}{k_B T}\right). \qquad (10.1.19)$$

Symmetry properties of the transition probabilities (10.1.16), (10.1.17) in spin variables (Ising kinetics) allow one to obtain the following expressions for the coefficients a_k, d_k (Asada 1990)

$$a_k = -d_0 d_k \quad (k \geq 1), \qquad (10.1.20)$$

$$\begin{aligned} d_1 &= (e^{\gamma} - 1)(1 + d_0 e^{\gamma})^{-1}, \\ d_2 &= (e^{\gamma} - 1)(1 + d_0 e^{2\gamma})^{-1} - 2d_1, \\ d_3 &= (e^{\gamma} - 1)(1 + d_0 e^{3\gamma})^{-1} - 3d_1 - 3d_2, \\ &\ldots\ldots \end{aligned} \qquad (10.1.21)$$

with

$$\gamma = \omega_2 / k_B T. \qquad (10.1.22)$$

In case of the Langmuir kinetics, where adsorption into cell j does not depend on the coverage of neighboring cells, in contrast to the desorption that does depend on this coverage, the following approximation is valid

$$a_k = 0, \quad d_k = (e^{\gamma} - 1)^k, \quad k \geq 1. \qquad (10.1.23)$$

If both adsorption and desorption depend symmetrically on the interaction with the nearest neighbors (kinetics with interaction), a_k and d_k look like

$$\begin{aligned} a_k &= -d_k, \quad k \geq 1, \\ d_1 &= \tanh\gamma, \\ d_2 &= \tanh 2\gamma - 2\tanh\gamma, \\ d_3 &= \tanh 3\gamma - 3\tanh 2\gamma + 3\tanh\gamma, \end{aligned} \qquad (10.1.24)$$

In asymmetrical case one has

$$a_k = -\alpha d_k, \quad k \geq 1, \qquad (10.1.25)$$

where α is a numerical constant of the order of unity.

For the Langmuir diffusion

$$D'_k = 0, \quad D_k = (e^{\gamma} - 1)^k, \quad k \geq 1, \qquad (10.1.26)$$

while for the case of kinetics with interaction

$$D_k = -\alpha D'_k, \quad D'_k = d_k, \quad k \geq 1. \qquad (10.1.27)$$

Taking into account (10.1.16)–(10.1.18) one can put down the master equation (10.1.2) as

$$\partial_t F(n,t) = \mathsf{M}_*(F) + \mathsf{AD}_*(F), \qquad (10.1.28)$$

where the operators M_*, AD_* describe the contributions of the diffusion and adsorption-desorption processes and have the form

$$\mathsf{M}_*(F) = \sum_{\bar{n}'} \sum_{j,k} \left[W_{Dj,j+k}(n',n)F(n',t) - W_{Dj,j+k}(n,n')F(n,t)\right], \qquad (10.1.29)$$

$$\begin{aligned}\mathsf{AD}_*(F) &= \mathsf{A}_*(F) - \mathsf{D}_*(F)\\ &= \sum_{\bar{n}'}\sum_{j}\left[W_{aj}(n',n)F(n',t) - W_{dj}(n,n')F(n,t)\right]. \qquad (10.1.30)\end{aligned}$$

In these relations $\bar{n}'$ means the configuration with the cells entering the internal sums (i.e those with numbers $j, j+k$) being absent. Equation (10.1.28), in distinction from (8.3.7)–(8.3.9), is the equation for the many-particle distribution function and therefore includes correlation effects. Thus, it can be used as a basis for the derivation of different models of correlation effects description. In particular, multiplying (10.1.28) by n_j, $n_j n_k$, $n_j n_k n_l$, ... and summing up over n gives (using definitions (10.1.12), (10.1.14)) the following system of coupled equations for one-, two-, etc. particle distribution functions:

$$\begin{aligned}\partial_t\theta_j &= \mathsf{M}_{*j}(\theta_j, \theta_{jk}) + \mathsf{AD}_{*j}(\theta_j, \theta_{jk}),\\ \partial_t\theta_{jk} &= \mathsf{M}_{*jk}(\theta_{jk}, \theta_{jkl}) + \mathsf{AD}_{*jk}(\theta_{jk}, \theta_{jkl}), \qquad (10.1.31)\\ &\ldots\ldots\end{aligned}$$

with

$$\mathsf{M}_{*j} = \sum_n n_j \mathsf{M}_*(F), \quad \mathsf{AD}_{*j} = \sum_n n_j \mathsf{AD}_*(F); \qquad (10.1.32)$$

$$\mathsf{M}_{*jk} = \sum_n n_j n_k \mathsf{M}_*(F), \quad \mathsf{AD}_{*jk} = \sum_n n_j n_k \mathsf{AD}_*(F). \qquad (10.1.33)$$

$$\ldots\ldots$$

Various approximations are based on different ways of decoupling the system (10.1.31). Different methods of the kinetic theory of gases can be used for that purpose, (Akhiezer and Peletminskiy 1977), (Dubrovskiy and Bogdanov 1979b) in particular.

10.2 Description of Adlayer Dynamics by Methods of Thermodynamics of Irreversible Processes

Dealing with isothermal concentrational kinetics one needs much less detailed information on the gas-surface interaction when using methods of the thermodynamics of irreversible processes (Keizer 1987). An advantage of these methods is in the general nature of basic assumptions and in handling only thermodynamical relations and interphase interaction parameters. But as for the Onsager coefficients, used in these methods, one has to appeal to experiment, to phenomenology or dynamical models of elementary processes.

Let us discuss main ideas of the above mentioned methods on an example of thin film growth kinetics for structureless gas flux under the conditions when diffusion processes are faster than those of adsorption-desorption (see. (8.3.7)). On the initial stage of film formation a role of diffusion is restricted to the promotion of two-dimensional nucleation (two–dimensional vapor of adatoms is assumed to be supersaturated). The study of growth kinetics at all stages of film formation can be done on the base of (8.3.7). On the other hand, if one is interested in the film growth description on large time scale (when nuclei are large and can be treated thermodynamically) methods of linear thermodynamics of irreversible processes can be used to solve the problem of nuclei evolution in a monolayer film (quasichemical approximation).

We shall characterize the system "gas + two-dimensional adsorbate" by the following densities: $n_3(t)$ – particle density in the gas phase; $n_1(t)$ – particle density in the condensed phase at the substrate; $n_2(t)$ – particle density in the two-dimensional gas; $n_{2'}(t)$ – particle density in the two-dimensional gas phase at an island surface (see Fig. 6). The picture, shown in Fig. 6, can be realized, for example, when one deals with the mechanism of chemisorption (phase 1) with internal (phase 2) and external (phase 2′) precursors (Lundquist et al. 1979), (Brivio and Grimley 1979), (Nørskov and Stoltze 1987), (Ceyer 1988).

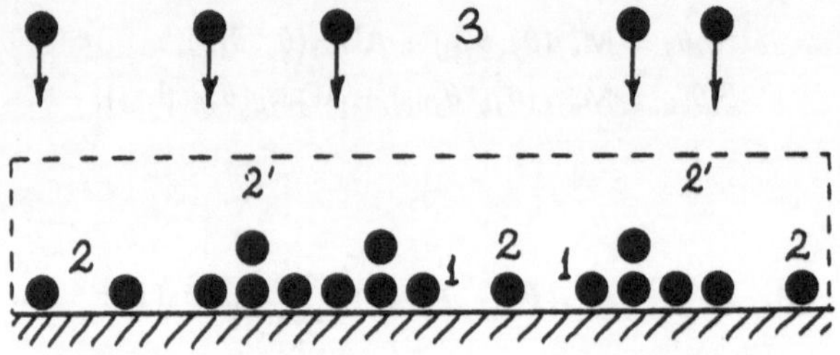

Fig. 6. Thermodynamical description of multiphase adsorbate dynamics: 1 – condensed phase (new phase nuclei at the substrate), 2 – two-dimensional gas at the substrate, 2′ – two-dimensional gas at the condensed phase surface, 3 – three-dimensional gas phase.

Let $N_3(t)$, $N_1(t)$, $N_2(t)$, $N_{2'}(t)$ be the total numbers of particles in the corresponding phases; S_1, S_2, S ($S = S_1 + S_2$) – the areas of islands, substrate covered by the two-dimensional gas, and total surface, respectively; V – the volume of the gas phase; N_c – the total number of adsorption cells at the surface; $n_c = N_c/S$ – the adsorption cell density. The coverages of different subsystems are

$$\theta_j(t) = N_j/N_c \quad (j = 1, 2, 2'). \tag{10.2.1}$$

The thermodynamical kinetics equations can be then written in the form

$$\dot{\theta}_j(t) = -\sum_k \frac{L_{jk}}{T}(\mu_k - \mu_j), \tag{10.2.2}$$

where = μ_j is the chemical potential of the j-th subsystem, L_{jk} are the Onsager coefficients ($L_{jk} = L_{kj}$).

An advantage of the equations (10.2.2) is that they include explicitly the conditions of thermodynamical equilibrium

$$\mu_k = \mu_j, \tag{10.2.3}$$

and that the chemical potentials can be calculated by the methods of equilibrium thermodynamics where the effects of non-ideal adsorbate can be included. Their shortage is in the uncertainty of the Onsager coefficients that are to be extracted from experiments, phenomenological or dynamical models. An explicit form of (10.2.2) for a specific system depends upon the approaches used for evaluation of the chemical potentials μ_j and the Onsager coefficients L_{ij}.

10.2.1 Ideal Two– and Three–Dimensional Gases

Suppose that the subsystems 3; 2, 2′ are ideal three- and two-dimensional gases. The normal bound energies of adatoms in systems 2 and 2′ (adsorption potentials) will be denoted as ε_2 and $\varepsilon_{2'}$ ($\varepsilon_2, \varepsilon_2' < 0$). Then the chemical potentials of the subsystems 3 and 2 are

$$\mu_3 = k_{\mathrm{B}}T \ln(\lambda^3 n_3), \tag{10.2.4}$$

$$\mu_2 = k_{\mathrm{B}}T \ln\left[\theta_2 \frac{\lambda^2 n_3}{Z_{\mathrm{n}}} \exp(-\frac{\varepsilon_2}{k_{\mathrm{B}}T})\right], \tag{10.2.5}$$

$$\lambda = \frac{h}{\sqrt{2\pi m k_{\mathrm{B}}T}} = hZ_0,$$

$$Z_{\mathrm{n}} = \exp\left(\frac{h\nu_{\mathrm{n}}}{2k_{\mathrm{B}}T}\right)\left[\exp\left(\frac{h\nu_{\mathrm{n}}}{k_{\mathrm{B}}T}\right) - 1\right]^{-1}.$$

Here λ is the thermal de Broglie wavelength, h - the Planck constant, Z_{n} and ν_{n} - the partition function and vibration frequency of adatom in normal direction. The chemical potential of condensed phase 1 will be calculated in approximation based on a simple Einstein model for liquid and solid condensate

$$\mu_1 = -q_{\mathrm{d}} + h\nu_{\mathrm{t}} + k_{\mathrm{B}}T \ln\left[\frac{\exp(-\frac{\varepsilon_2}{k_{\mathrm{B}}T})}{Z_{\mathrm{n}}Z_{\mathrm{t}}}\right], \tag{10.2.6}$$

$$Z_{\mathrm{t}} = \exp\left(\frac{h\nu_{\mathrm{t}}}{k_{\mathrm{B}}T}\right)\left[\exp(\frac{h\nu_{\mathrm{t}}}{k_{\mathrm{B}}T}) - 1\right]^{-1},$$

where q_{d} is the heat of evaporation of a droplet, Z_{t} and ν_{t} are the partition function and vibration frequency of adatom inside an island in tangential direction. It is assumed that islands do not coagulate and have arbitrary shaped flat configurations. Relevant adatom fluxes are calculated in the framework of free molecular or jumping mechanism of deposition. It is further assumed that the frequency of adsorption-desorption processes is much greater than that of migration of adatoms over the surface, therefore the coverage can be considered to be constant all over the surface.

The ideal gas equations of state for three- and two-dimensional subsystems have the form

$$P = n_3 k_{\mathrm{B}}T; \quad \Pi_i = n_i k_{\mathrm{B}}T, \quad i = 2, 2', \tag{10.2.7}$$

where P and Π are the bulk and surface gas pressures, respectively. The experimental laws (Knudsen, Henry) can be used to specify the Onsager coefficients. Then the system (10.2.1) can be written in the explicit form

$$\dot{\theta}_j(t) = \sum_{k \neq j} r_{jk}(\theta_l), \quad j = 1, 2, 2'; \quad k = 1, 2, 2', 3. \tag{10.2.8}$$

The evaluation of $r_{jk}(\theta_l)$ on the base of (10.2.1) and accepted models of the phases results in (Kreuzer and Payne 1988a, 1988b, 1989)

$$r_{12} = a_* y_0 \left(\frac{\theta_1(0) n_c}{N_0}\right)^{1-\zeta} \left(\frac{\theta_1}{\theta_{1c}}\right)^{\zeta} k_B T Z_0 \left(\frac{\theta_2}{1-\theta_1/\theta_{1c}} - B_1 \Theta(\theta_1)\right),$$

$$r_{12'} = a'_* y'_0 \left(\frac{\theta_1(0) n_c}{N_0}\right)^{1-\zeta'} \left(\frac{\theta_1}{\theta_{1c}}\right)^{\zeta'} \left(\frac{\theta_{2'}\theta_{1c}}{\theta_1} - B_0 B_1 \Theta(\theta_1)\right),$$

$$r_{13} = \begin{cases} \frac{a_{1*}}{n_c}\frac{\theta_1}{\theta_{1c}}\left[j - \nu_1 n_c B_1 \exp\left(\frac{\varepsilon_2}{k_B T}\right)\right], & \theta_1 \le \theta_{1c} \\ \frac{a_{1*}}{n_c}\left[j - \nu_1 n_c \exp\left(-\frac{q_d - \varepsilon_2}{k_B T}\right)\right], & \theta_1 > \theta_{1c} \end{cases}, \tag{10.2.9}$$

$$r_{23} = a_{2*}\left[\frac{1}{n_c}(1 - \frac{\theta_1}{\theta_{1c}})j - \nu_1 \exp(-\frac{\varepsilon_2}{k_B T})\theta_2\right],$$

$$r_{2'3} = a_{2'*}\left[\frac{1}{n_c}\frac{\theta_1}{\theta_{1c}} j - \nu'_1 \exp(-\frac{\varepsilon_{2'}}{k_B T})\theta_{2'}\right],$$

$$r_{21} = -r_{12}, \quad r_{22'} = \frac{1 - a'_*}{a'_*} r_{12}, \quad r_{2'1} = -r_{12'}, \quad r_{2'2} = -r_{22'},$$

$$j = P Z_0, \quad \theta_{1c} = \frac{n_1}{n_c}, \quad \frac{\theta_1}{\theta_{1c}} = \frac{S_1}{S},$$

$$B_1 = \frac{1}{\lambda^2 n_c}\left[1 - \exp\left(-\frac{h\nu_t}{k_B T}\right)\right]^2 \exp\left(-\frac{q_d}{k_B T}\right),$$

$$B_0 = \frac{\nu_1}{\nu'_1}\exp\left(\frac{\varepsilon_2 - \varepsilon_{2'}}{k_B T}\right), \quad \Theta(\theta_1) = \begin{cases} 1, & \theta_1 > 0 \\ 0, & \theta_1 = 0 \end{cases}.$$

In the expressions above n_1 is adatom density in the subsystem 1, θ_{1c} is the substrate coverage by the phase 1 in the upper point of the co-existence curve (if $\theta_1 \le \theta_{1c}$ two phases co-exist, if $\theta_1 > \theta_{1c}$ only the condensed phase exists, when $\theta_2 \approx 0, \theta_1/\theta_{1c} \approx 1$, $S_2 \approx 0$, $S_1 \approx S_2 \approx S$), N_0 is the mean number of adatoms per one island, a_* and a'_* are the rates of adatom capture by an island from the adsorbate at the substrate and that at the surface of the island, a_{1*}, a_{2*}, and $a_{2'*}$ are the probabilities of adatom adsorption into the phases 1, 2, and 2′, respectively.

A phenomenological equation of islands growth through adatoms capture

$$\dot{n}_1 = a_* S_1^{\zeta} y \Delta j_{\tau}, \tag{10.2.10}$$

has been used to obtain (10.2.9). The general equation for adatom flux density at the surface has the form

$$\Delta j_{\tau} = (n_2 - \bar{n}_2)\bar{v}; \quad (n_2 - \bar{n}_2) = (k_B T)^{-1}(\Pi_2 - \bar{\Pi}_2). \tag{10.2.11}$$

Different expressions for the mean velocity $\bar{v}$ of adatom migration over the substrate can be used. Thus, for movable and localized adsorbate one has respectively

$$\bar{v} = Z_0/2\mu, \tag{10.2.12}$$

and

$$\bar{v} = \frac{l_d}{\tau_D}\exp\left(-\frac{\varepsilon_D}{k_B T}\right), \tag{10.2.13}$$

where l_d is the lattice constant, τ_D^{-1} is the frequency of diffusion jumps, ε_D is the height of the diffusion barrier. Quantities y and ζ depend on the form of flat island and are coupled by the relations

$$y = y_0 \tilde{n}^{1-\zeta}, \quad y_0 \geq 2\sqrt{\pi}, \quad 1/2 \leq \zeta < 1, \tag{10.2.14}$$

with $\tilde{n}$ being the numerical density of islands. For the model of circular disks, for example, one has

$$\zeta = 1/2, \quad y_0 = 2\sqrt{\pi}. \tag{10.2.15}$$

One can prove that the condition of the subsystem 2 ideality is

$$B_1 \leq 1. \tag{10.2.16}$$

If (10.2.16) is not met, one has to include non–ideal effects in the two–dimensional phase description. It is worth pointing that in the model of ideal two–dimensional gas the lower point of phases co-existence for the phases 1 and 2 (θ_{2c}) is not defined at all, while the upper one (θ_{1c}) is introduced only formally. In the models of non-ideal gas these points can be determined from the coverage versus the chemical potential curve.

10.2.2 Non–Ideal Two–Dimensional Gas Models

The following analytical expressions for the chemical potential of two–dimensional non–ideal gas are known

Van-der-Waals approximation:

$$\mu = k_{\mathrm{B}}T \ln \left[\frac{\theta}{1-\theta} \frac{\lambda^2 n_{\mathrm{c}}}{Z_{\mathrm{n}}} \exp\left(\frac{\varepsilon_2}{k_{\mathrm{B}}T}\right)\right] - 2\alpha n_{\mathrm{c}}\theta + k_{\mathrm{B}}T \frac{\theta}{1-\theta}, \tag{10.2.17}$$

where $\alpha = \omega_2/2n_{\mathrm{c}}$ is a pressure parameter, ω_2 is the interaction energy of the nearest neighbors.

Bragg–Williams approximation:

$$\mu = k_{\mathrm{B}}T \ln \left[\frac{\theta}{1-\theta} \frac{1}{Z_{\mathrm{n}} Z_{\mathrm{t}}}\right] \exp\left(\frac{\varepsilon_2}{k_{\mathrm{B}}T}\right) - \kappa\omega_2\theta, \tag{10.2.18}$$

where κ is the coordination number.

Quasichemical LG model:

$$\mu = k_{\mathrm{B}}T \ln \left[\left(\frac{\theta}{1-\theta} \frac{1}{Z_{\mathrm{n}} Z_{\mathrm{t}}}\right) \exp\left(\frac{\varepsilon_2}{k_{\mathrm{B}}T}\right)\right] - \frac{\kappa}{2}\omega_2$$

$$+ \frac{\kappa}{2} k_{\mathrm{B}}T \ln \frac{\gamma - 1 + 2\theta}{\gamma + 1 - 2\theta}, \tag{10.2.19}$$

$$\gamma^2 = 1 - 4\theta(1-\theta)\left[1 - \exp\left(\frac{\omega_2}{k_{\mathrm{B}}T}\right)\right]. \tag{10.2.20}$$

Using these expressions one can derive corresponding equations of state of the two-dimensional gas. For example, from (10.2.17) one can get

$$\Pi(\theta) = n_{\mathrm{c}} k_{\mathrm{B}}T \frac{\theta}{1-\theta} - \alpha n_{\mathrm{c}}^2 \theta^2. \tag{10.2.21}$$

When S and T are fixed, the lower (θ_{2c}) and the upper (θ_{1c}) points of the phases co-existence are to be found from conditions

$$\Pi(\theta_{2c}, S, T) = \Pi(\theta_{1c}, S, T), \tag{10.2.22}$$

$$\mu(\theta_{2c}, S, T) = \mu(\theta_{1c}, S, T). \tag{10.2.23}$$

For given θ the equilibrium sharing ($\bar{\theta}_2$ and $\bar{\theta}_1$) of adatoms between the phases is found from the Maxwell rule (Kashchiev 1976) resulting in

$$\bar{\theta}_2 = \theta_{2c}\frac{\theta_{1c} - \theta}{\theta_{1c} - \theta_{2c}}; \quad \bar{\theta}_1 = \theta_{1c}\frac{\theta - \theta_{2c}}{\theta_{1c} - \theta_{2c}}; \quad \bar{\theta}_1 + \bar{\theta}_2 = \theta, \tag{10.2.24}$$

and

$$\bar{S}_2 = \frac{\bar{\theta}_2}{\theta_{2c}}S; \quad \bar{S}_1 = \frac{\bar{\theta}_1}{\theta_{1c}}S; \quad \bar{S}_1 + \bar{S}_2 = S. \tag{10.2.25}$$

The chemical potential of the phase 1 is constant for any sharing between the phases

$$\mu_1 = \mu_1(\theta_{1c}) = \mu_{1c}. \tag{10.2.26}$$

The chemical potential of the phase 2 depends on its coverage, i.e.

$$\mu_2 = \mu_2(\theta_2). \tag{10.2.27}$$

When $\mu_2 > \mu_1$, $\theta_2 > \theta_{2c}$ and the two-dimensional gas partially condenses, when $\mu_2 < \mu_1$ $\theta_2 < \theta_{2c}$ and a part of condensate evaporates into the two-dimensional gas.

For the Bragg-Williams model the quantities θ_{jc}, μ_{jc} are obtained from relations

$$\ln\left(\frac{\theta_{jc}}{1 - \theta_{jc}}\right) - \frac{\kappa}{2}\omega_2 - k_{\mathrm{B}}T\ln\left[Z_{\mathrm{n}}Z_{\mathrm{t}}\exp\left(\frac{\varepsilon_2}{k_{\mathrm{B}}T}\right)\right] = 0, \quad j = 1, 2, \tag{10.2.28}$$

$$\theta_{2c} + \theta_{1c} = \theta, \tag{10.2.29}$$

$$\mu_{2c} = \mu_{1c} = -\frac{\kappa}{2}\omega_1 - k_{\mathrm{B}}T\ln\left[Z_{\mathrm{n}}Z_{\mathrm{t}}\exp\left(\frac{\varepsilon_2}{k_{\mathrm{B}}T}\right)\right]. \tag{10.2.30}$$

Let us assume now that all the phases at the substrate are in equilibrium, i.e.

$$\mu_1(\theta, S, T) = \mu_2(\theta, S, T) = \mu_{2'}(\theta, S, T) \equiv \mu(\theta, S, T), \tag{10.2.31}$$

$$\theta = \theta_1 + \theta_2 + \theta_{2'}, \tag{10.2.32}$$

and consider the quasiequilibrium adsorption model ($\mu_j \neq \mu_3$). The kinetic equation for the layer coverage θ can then be presented in a form

$$\dot{\theta} = -\sum_{j=1,2,2'}\frac{L_{j3}}{T}(\mu_3 - \mu_j). \tag{10.2.33}$$

This equation is equivalent to that of a "point" model obtained from (8.3.7)

$$\dot{\theta} = \mathsf{AD}_*(\theta) \tag{10.2.34}$$

and can be transformed to the form

$$\dot{\theta} = [(a_{1t*} - a_{2*})\theta + a_{2*}\theta_{1c} - a_{1*}\theta_{2c}] \frac{1}{(\theta_{1c} - \theta_{2c})} \frac{\lambda^3 P}{k_B T n_c}, \qquad (10.2.35)$$

$$a_{1t*} = a_{1*} + a_{2'*}. \qquad (10.2.36)$$

Thus, the kinetic equation of the quasiequilibrium adsorbate model involves a set of adsorption coefficients, co-existence parameters θ_{2c}, θ_{1c} and the gas phase parameters T, P.

Let us consider now more detailed kinetics, described by the quasichemical model in the Bragg–Williams approximation. We shall use an expression for the chemical potential of the phase 2 that takes into account the time dependence of the area covered with this phase

$$\mu_2 = k_B T \ln \left[\frac{\theta_2}{\theta_1' - \theta_2} \frac{1}{Z_n Z_t} \exp\left(\frac{\varepsilon_2}{k_B T}\right) \right] - \kappa\omega_2 \frac{\theta_2}{\theta_1'}, \qquad (10.2.37)$$

$$\theta_1' = 1 - \frac{\theta_1}{\theta_{1c}}. \qquad (10.2.38)$$

The partial coverages θ_1, θ_2 are not equal to the equilibrium values $\bar{\theta}_1$, $\bar{\theta}_2$, but the following relation holds

$$\theta_1(t) + \theta_2(t) = \bar{\theta}_1(t) + \bar{\theta}_2(t) = \theta(t). \qquad (10.2.39)$$

Far from equilibrium in the upper point $\theta_1 \gg \bar{\theta}_1$, $\theta_2 \ll \bar{\theta}_2$, while in the lower one $\theta_1 \ll \bar{\theta}_1$, $\theta_2 \gg \bar{\theta}_2$. The chemical potential of the subsystem $2'$ may be taken from the ideal LG approximation

$$\mu_{2'} = k_B T \ln \left[\frac{\theta_{2'}}{\Theta_1} \frac{1}{Z_n' Z_t'} \exp\left(\frac{\varepsilon_{2'}}{k_B T}\right) \right], \qquad (10.2.40)$$

$$\Theta_1 = \begin{cases} \theta_1/\theta_{1c}, & 0 < \theta_1 \le \theta_{1c} \\ 1, & \theta_1 > \theta_{1c} \end{cases}. \qquad (10.2.41)$$

The equilibrium occupancies are defined by

$$\bar{\theta}_1 = \theta_{1c} \frac{\theta - \theta_{2c}}{Y + \theta_{1c} + \theta_{2c}}; \quad \bar{\theta}_2 = \theta - \theta_{1c} - \theta_{2c}; \quad \bar{\theta}_{2'} = \bar{\theta}_1/\theta_{1c} Y, \qquad (10.2.42)$$

$$Y = \frac{Z_n' Z_t'}{Z_n Z_t} \exp\left[\frac{1}{k_B T}(-\varepsilon_{2'} + \varepsilon_2 + \omega_2(1 - \frac{\kappa}{2}))\right].$$

Kinetic equations for the partial coverages just as above have the form (10.2.8). Quantities r_{jk} (excluding the terms describing adsorption ($P = 0$), that keep unchanged) are as follows.

Above the phases co-existence point ($\theta_1 > \theta_{1c}$, $\theta_2 \approx 0$)

$$r_{12} = r_{22'} = r_{12'} = r_{23} = 0,$$

$$r_{13} = -a_{1*}\Phi \frac{\theta_1}{1 - \theta_1} \exp\left[-\frac{1}{k_B T}(\kappa\omega_2\theta_1 - \varepsilon_2)\right], \qquad (10.2.43)$$

$$r_{2'3} = -a_{2'*}\Phi' \theta_{2'} \exp\left[-\frac{1}{k_B T}(\omega_2 - \varepsilon_{2'})\right],$$

$$\Phi = k_B T (h Z_n Z_t n_c \lambda^2)^{-1}, \quad \Phi' = k_B T (h Z_n' Z_t' n_c \lambda^2)^{-1}.$$

In the region of the phases co-existence ($\theta_1' = 1 - \theta_1/\theta_{1c} > 0$)

$$
\begin{aligned}
r_{13} &= -a_{1*}\Phi\frac{\theta_1}{\theta_{1c}}\exp\left(-\frac{\kappa\omega_2}{2k_BT}\right),\\
r_{23} &= -a_{2*}\Phi\frac{\theta_1'\theta_2}{\theta_1' - \theta_2}\exp\left[-\frac{1}{k_BT}(\kappa\omega_2\theta_2/\theta_1' - \varepsilon_2)\right],\\
r_{2'3} &= -a_{2'*}\Phi'\theta_{2'}\exp\left(-\frac{\omega_2 + \varepsilon_{2'}}{k_BT}\right),\\
r_{12} &= a_*\Gamma(\zeta, N_0)Z_0k_BT\exp\left(\frac{\varepsilon_1}{k_BT}\right)\frac{\theta_2 - \bar{\theta}_2}{\theta_1'},\\
r_{12'} &= a_*'\Gamma(\zeta', N_0)Z_0k_BT\exp\left(\frac{\varepsilon_1}{k_BT}\right)\frac{\theta_{1c}}{\theta_1}(\theta_{2'} - \bar{\theta}_{2'}),\\
r_{22'} &= \frac{1 - a_* - b_*}{a_*}r_{12}, \quad yS_1^\zeta = S_1\Gamma(\zeta, N_0),
\end{aligned}
\tag{10.2.44}
$$

where b_* is the probability of adatom reflection from an island boundary.

Under the phases co-existence point ($\theta_1 = 0$)

$$
\begin{aligned}
r_{jk} &= 0 \quad (j \neq 2, k = 3),\\
r_{23} &= -a_{2*}\Phi\frac{\theta_2}{1 - \theta_2}\exp\left[-\frac{1}{k_BT}(\kappa\omega_2\theta_2 - varepsilon_2)\right].
\end{aligned}
\tag{10.2.45}
$$

10.2.3 Non–Equilibrium Thermodynamics of Diffusion

Here we shall review an application of methods of non–equilibrium thermodynamics to diffusion processes description (Murch and Thorn 1979). The following three types of particles are assumed to exist at the surface: adatoms beyond clusters (particles of the sort 2 that form a gas with a small gradient of the chemical potential μ_2); condensed adatoms (particles of the sort 1 that form a media with a large gradient of μ_1); and substrate vacancies (particles of sort 3). Diffusion fluxes of particles of the sort $j = 1, 2, 3$ can be written by analogy with (10.2.2):

$$I_j = -\sum_k \frac{L_{jk}}{T}\partial_{\boldsymbol{R}}\mu_k(\boldsymbol{R}, t), \tag{10.2.46}$$

$$\theta_1 + \theta_2 + \theta_3 = 1, \quad I_1 + I_2 + I_3 = 0; \quad \theta_j = N_c n_j. \tag{10.2.47}$$

We shall assume, at first, the unified subsystem (1+2) to be in equilibrium with the system 3. It means that $I_3 = 0$, $\theta_1 + \theta_2 = 1$, and diffusion results in a redistribution of adatoms between the systems 2 and 1. Therefore

$$\partial_{\boldsymbol{R}}\mu_j = \frac{k_BT}{\theta_j}\partial_{\boldsymbol{R}}\theta_j \quad (j = 1, 2), \quad \partial_{\boldsymbol{R}}\theta_1 = -\partial_{\boldsymbol{R}}\theta_2; \tag{10.2.48}$$

$$I_2 = D_{*\mathrm{tr}}^R\partial_{\boldsymbol{R}}\theta_2(\boldsymbol{R}, t), \quad I_1 = -D_{*\mathrm{tr}}^R\partial_{\boldsymbol{R}}\theta_1(\boldsymbol{R}, t); \tag{10.2.49}$$

$$D_{*\mathrm{tr}}^R = k_B\left(\frac{L_{21}}{\theta_1} - \frac{L_{22}}{\theta_2}\right). \tag{10.2.50}$$

The Onsager coefficients L_{jk} ($j, k = 1, 2$) meet the following relations

$$\theta_1^{-1}(L_{11}+L_{21})=\theta_2^{-1}(L_{12}+L_{22}). \tag{10.2.51}$$

The diffusion coefficient $D^R_{*\mathrm{tr}}$ is known as a trajectory diffusion coefficient (or self-diffusion coefficient).

Coming to the general case, let us introduce the total flux (I_a) of the particles of the sorts 1 and 2

$$I_a=I_1+I_2=-I_3=-\sum_{j=1}^{2}\sum_{k=1}^{3}L_{jk}\partial_{\boldsymbol{R}}\mu_k, \tag{10.2.52}$$

$$L_{1k}+L_{2k}+L_{3k}=0 \quad (k=1,2). \tag{10.2.53}$$

Using the Gibbs–Dughuem relation

$$\sum_{j=1}^{3}\theta_j\partial_{\boldsymbol{R}}\mu_j=0, \tag{10.2.54}$$

the normalizing condition (10.2.47), and expression for the chemical potential of vacancies

$$\mu_3=\mu_3^0+k_{\mathrm{B}}T\ln(\theta_3 Z_3), \tag{10.2.55}$$

$$Z_3=\exp\left(\frac{\varepsilon_a}{k_{\mathrm{B}}T}\right), \tag{10.2.56}$$

with Z_3 being a vacancy activity coefficient, yields for the total diffusion flux of adsorbed particles

$$I_a=-D^R_{*\mathrm{ch}}\partial_{\boldsymbol{R}}\theta_a, \tag{10.2.57}$$

where

$$D^R_{*\mathrm{ch}}=D^R_{*\mathrm{tr}}\left[\theta_3^{-1}+\frac{k_{\mathrm{B}}}{N}\frac{L_{21}}{D^R_{*tr}}\frac{(1-\theta_1)}{\theta_1\theta_2\theta_3}\right](1+\theta_3\frac{\partial Z_3}{\partial\theta_3}) \tag{10.2.58}$$

Diffusion equations have the form

$$\begin{aligned}\partial_t\theta_2(\boldsymbol{R},t)&=-\partial_{\boldsymbol{R}}I_2,\\ \partial_t\theta_1(\boldsymbol{R},t)&=-\partial_{\boldsymbol{R}}I_1,\\ \partial_t\theta_a(\boldsymbol{R},t)&=-\partial_{\boldsymbol{R}}I_a.\end{aligned}$$

The expressions (10.2.52), (10.2.59) contain unknown Onsager coefficients that are to be found from independent considerations.

10.2.4 Diffusion Coefficients

Let us discuss a question of analytical approximations for the diffusion coefficients.

For low coverages ($\theta\to 0$) one-dimensional LG model leads to the following expression for the diffusion coefficient (Kreuzer 1990) ($D^R_{*\mathrm{tr}}\equiv D^R_*$)

$$D^R_*(0)=\frac{1}{2}\nu_t l_s^2\exp\left(-\frac{\varepsilon_D}{k_{\mathrm{B}}T}\right), \tag{10.2.60}$$

where ν_t is the frequency of adatom vibrations in tangential direction, l_s is the lattice constant, ε_D is the diffusion activation barrier.

Lateral interaction can be included in the one-dimensional model through a representation

$$D_*^R(\theta) = \frac{1}{2}\nu_t < (1 - n_{j+1}) \exp\left(-\frac{E_{j,j+1}}{k_B T} \right) >_j, \quad (10.2.61)$$
$$< n_j >_j = \theta,$$

where symbol $< ... >_j$ denotes averaging over different lattice configurations, $E_{j,j+1}$ is the energy barrier for the transition of an adatom from the j-th cell to the $j+1$-st one. The following expression can be used for the latter quantity (see Fig. 7)

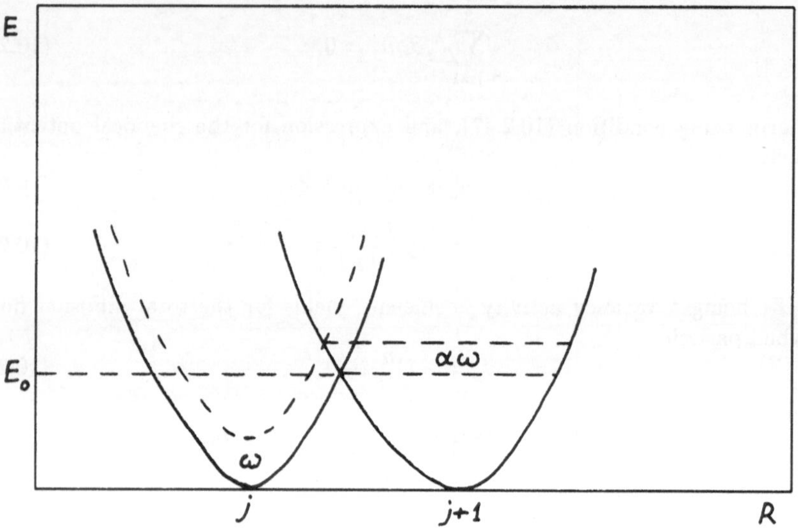

Fig. 7. To the calculation of the activation barrier of a particle jump $j \to j+1$ with an eye to lateral interactions.

$$E_{j,j+1} = E_0 - (1-\alpha)\omega_2 n_{j-2} + [-(1-\alpha)\omega_1 + \alpha\omega_2] n_{j-1} + \\ +[-(1-\alpha)\omega_2 + \alpha\omega_1] n_{j+2} + \alpha\omega_2 n_{j+3}, \quad (10.2.62)$$

where $\omega_1, \omega_2, \ldots$ are the interaction energies of adatom with the nearest neighbor, next by one neighbor, etc., α is a parameter ($0 \le \alpha \le 1$). Let $W\{n_k|n_l\}$ be the probability of realization of the configuration n_k under condition that l places, not contained in n_k, form the configuration n_l. Let $W\{\overline{j+1}|j\}$ be the probability of the $j+1$-st cell to be free, while the j-th cell is occupied. Then for the finite coverages expression (10.2.61) can be presented in the form

$$D_*^R(\theta) = D_*^R(0) W\{\overline{j+1}|j\} \exp\left(-\frac{\Delta E_*}{k_B T} \right), \quad (10.2.63)$$

with $\exp(-\Delta E_*/k_B T)$ being the thermodynamical mean value of $\exp(-\Delta E_{j,j+1}/k_B T)$ $(\Delta E_{j,j+1} = E_{j,j+1} - E_0)$:

$$\exp\left(-\frac{\Delta E_*}{k_B T}\right) = \sum_{\substack{n_{j-2},n_{j-1}\\ n_{j+2},n_{j+3}}} \exp\left(-\frac{\Delta E_{j,j+1}}{k_B T}\right) W\{n_{j-2}, n_{j-1}; n_{j+2}, n_{j+3}|j, \overline{j+1}\}.$$

Under equilibrium $W\{n_k|n\}$ can be written as

$$W\{n_k|n_l\} = Z(n_k|n_l)/Z(n_l),$$

where Z is the equilibrium partition function of lattice gas, calculated in the Bethe-Peierls approximation. The results of numerical calculations on the base of (10.2.63), presented in Fig. 8, demonstrate the influence of lateral interactions on the diffusion coefficient. For $\theta \to 0$ the diffusion coefficient tends to its classical limit D_0.

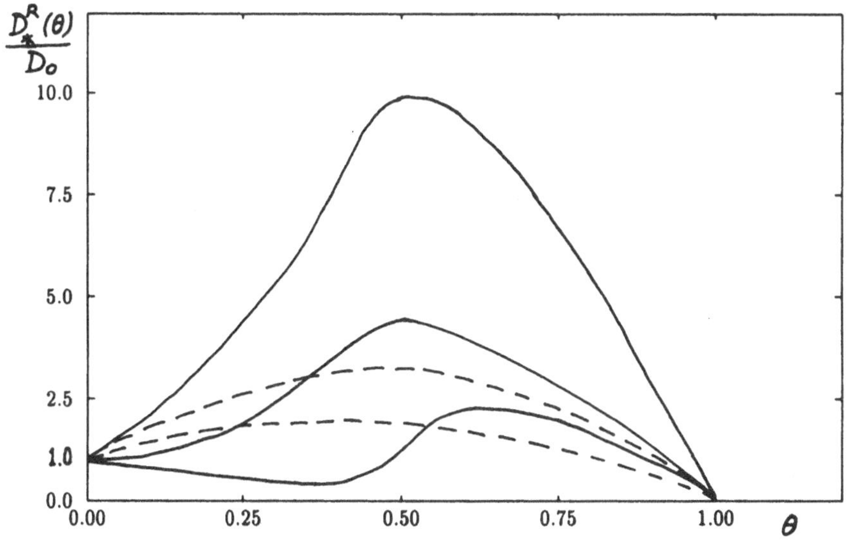

Fig. 8. Diffusion coefficient dependence on layer coverage at different values of $(\varepsilon/k_B T, \alpha)$: 1 - (0,0); 2 - (-1,0); 3 - (-2,0); 4 - (0,0.25); 5 - (0,0.5).

The model of two-dimensional cells (cellular model) leads to the following expressions for D_*^R, $D_{*\mathrm{ch}}^R$ (Reed and Ehrlich 1981, Zubcus and Tornau 1989):

$$D_*^R = \frac{3}{2} l_s^2 < R_1 >, \quad D_{*\mathrm{ch}}^R = D_*^R \left[\frac{1}{k_B T} \frac{\partial \mu}{\partial \theta}\bigg|_T \right], \tag{10.2.64}$$

where $< R_1 >$ is the velocity of an adatom jumping into a neighboring cell, averaged over all configurations of the lattice gas, $\mu(\theta)$ is the chemical potential of the two-dimensional gas. The following limits hold

$$D_*^R \to D_{*\mathrm{ch}}^R \to \frac{3}{2} l_s^2 R_{10} \quad (\theta \to 0, \mu \to -\infty), \tag{10.2.65}$$

$$D_*^R \to 0, \quad D_{*\mathrm{ch}}^R \to \frac{3}{2} l_s^2 R_1(0,1,1,\ldots) \quad (\theta \to 1, \mu \to +\infty), \tag{10.2.66}$$

with R_{10} being the velocity of an adatom jumping into neighboring cell at the free substrate and $R_1(0,1,1\ldots)$ is that of jumping into the nearest free cell under the condition that all the rest neighboring cells are occupied. Particularly, for the Langmuir monolayer one has

$$D_*^R = \frac{3}{2} l_s^2 R_{10}(1-\theta), \quad D_{*\mathrm{ch}}^R = \frac{3}{2} l_s^2 R_{10}. \tag{10.2.67}$$

In (Ruzaykin and Ervye 1989) the following analytical approximations for the concentration dependence of the surface diffusion coefficient (for a lattice gas without lateral interactions) have been proposed

$$D_*^R(\theta) = \frac{\nu_t l_s^2}{12}[3 + 8(\theta + \theta^2 + \theta^3) - 15\theta^4] \tag{10.2.68}$$

for a square lattice, and

$$D_*^R(\theta) = \frac{\nu_t l_s^2}{6}[2 + 9\theta - 5\theta^2] \tag{10.2.69}$$

for a hexagonal lattice with the atoms in its nodes. Including lateral interactions via the nearest neighbors approximation yields instead of (10.2.68), (10.2.69)

$$D_*^R(\theta) = \frac{\nu_t l_s^2}{12}[3 + 8(\theta + \theta^2 + \theta^3) - 15\theta^4]\left\{1 + \frac{2\theta(\delta + 1 - 2\theta)}{\delta - 1 + 2\theta}\right) \\ \times \left[4(1-\theta)^2\left(\frac{1}{\delta}(1 - Z_3)(2\theta - 1) - Z_3\right) + \delta(5 - 6\theta) + 1 + 4\theta^2\right]\right\} \tag{10.2.70}$$

for a square lattice, and

$$D_*^R = \frac{\nu_t l_s^2}{6}[2 + 9\theta - 5\theta^2]\left\{1 + \frac{3}{2}\frac{\theta(\delta + 1 - 2\theta)}{\delta - 1 + 2\theta} \\ \times \left[4(1-\theta)^2\left(\frac{1}{\delta}(1 - Z_3(2\theta - 1) - Z_3\right) + \delta(5 - 6\theta) + 1 + 4\theta^2\right]\right\} \tag{10.2.71}$$

for a hexagonal lattice. In (10.2.70) and (10.2.71)

$$\delta^2 = 1 - 4\theta(1-\theta)(1 - Z_3), \quad Z_3 = \exp(-\varepsilon_a/k_B T). \tag{10.2.72}$$

In some papers (Pereira and Zgrablich 1989, Zgrablich et al. 1986) both surface inhomogeniety and lateral interactions influence on the diffusion coefficient have been studied. There has been considered the diffusion over an inhomogeneous energetic relief of the surface under the following simplifying assumptions:

1. For every pair of sites j and k separated by the distance R_{jk} the level $-\epsilon_{jk} \le -\epsilon_c$ exists that is minimal and coupling both places ($\epsilon_{jk} = \epsilon_{kj}$). Generally, the level ϵ_{jk} lies under the critical one ϵ_c. The latter defines the percolation threshold (see Fig. 9).

2. An adatom, jumping from the j-th place to a free k-th one should get into an activated state characterized by the activatin barrier $-\varepsilon_{jk}$ and the energy of interaction with other adatoms.
3. This interaction is described by the mean field energy $-\omega$ for adatoms in the ground state and by $-\omega^*$ for those in activated ones.
4. The probability of the occupation of an empty place with an adatom is 1. The probability of an adatom getting out to the gas phase (after the collision with an occupied place) is p_d and the probability to continue the diffusion is $1 - p_d$.

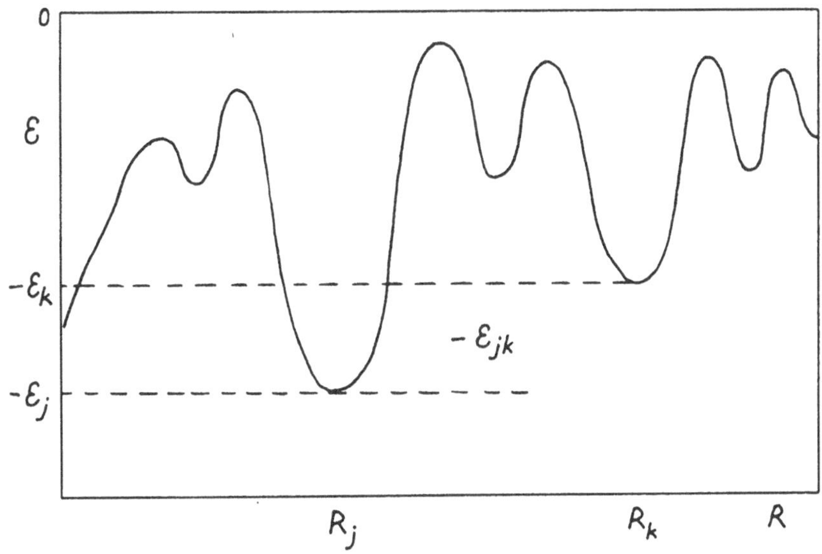

Fig. 9. Profile of potential relief for diffusing adatom.

Under the above assumptions the following expression for the diffusion coefficient has been obtained

$$D_*^R(\theta) = D_*^R(0)(1-\theta)^2 \frac{1}{k_B T} \exp\left[\frac{1}{k_B T}(\omega^*\theta + \mu)\right] \frac{\partial \mu}{\partial \theta} \exp\left(\frac{-T_0 \Phi^2(\theta)}{T}\right). \quad (10.2.73)$$

Here T_0 is the characteristic surface temperature

$$T_0 = \frac{\delta_0}{k_B \rho_0 \lambda_0^2} = \frac{2\delta_0}{k_B} \Delta_E, \quad \rho_0 = \frac{1}{2\Delta_E \lambda_0^2}, \quad (10.2.74)$$

where Δ_E is the dispersion of adsorption energy, i.e. the measure of the surface inhomogeniety, δ_0 is a quantity related to the percolation threshold ($\delta_0 \approx 0.16$) (Zallen 1988), λ_0 is the distance between the nearest adsorption sites, $\lambda_0/\Phi(\theta)$ is the mean free path of an adatom along the surface. The Monte - Carlo simulations have given the following approximation for $\Phi(\theta)$ (Zgrablich et al. 1986)

$$\Phi(\theta) = a_0(1-\theta) + a_1\theta, \quad (10.2.75)$$

where a_0 is the probability of adsorption to a free site, a_1 is the probability of desorption after the collision with an adatom in a cell. Thus, expression (10.2.75) includes both lateral interactions and physical inhomogeniety (i.e. a variation of the adsorption potential) of the surface. æ

III

Phenomenological Models of Thin Film Growth at Solid Surfaces

Introduction

Different phenomenological and thermodynamical models of thin film formation together with more rigorous approaches presented in the previous Part play a significant role in study of the film growth dynamics. An advantage of phenomenological and thermodymanical models is the description of growth in terms of "technological" parameters (flux density, gas phase pressure, gas and substrate temperatures, deposition regime, substrate state parameters, etc.). A shortage is in the absence of their first-principle substantiation, lack of generallity that leads to difficulties in comparison of the results of different models, existence of empirical (fitting) parameters. Nevertheless, phenomenological models provide a useful instrument for the study of physics of deposition and growth, represent adequatly mechanisms of these processes at specific growth stages and allow one to interpret experimental results.

Combining phenomenological and first-principle approaches gives one a possibility to study in detail the growth processes during all stages starting from elementary acts of interphase interactions. We do not aim to present here all the variety of existing phenomenological models, we shall rather present, firstly, the most important and most productive ones and, secondly, the approaches that are close to the methods discussed in Part II.

11 Classical Model of Nucleation and Nuclei Growth at Substrate

11.1 Incubation Stage

The film growth kinetics includes generally a hierarchy of a number of processes: adsorption-desorption, diffusion of single adatoms, cluster formation leading to the nucleation of critical nuclei (either liquid or solid), isolated nuclei growth, coalescence (recondensation), secondary nucleation, coagulation of nuclei, the growth of continuous film with "healing" nonuniformities and voids. The classical model of nucleation at substrate that describes the kinetics of liquid droplets formation from a two-dimensional vapor is based on the capillary model of nuclei growth (Osipov 1990). This model assumes that the following relations between characteristic times of the phase transition kinetics take place

$$t_W \ll t_G \ll t_C \tag{11.1.1}$$

with t_W being the time of formation of the quasistationary distribution of clusters over their sizes, t_G - the time of isolated growth of nuclei, and t_C - the time scale of liquid droplets coagulation. In addition, the following characteristic times are to be specified: mean life time of an adatom at the substarte t_a, characteristic deposition time t_j, characteristic diffusion time t_d, and the time of maximal supersaturation formation t_s.

At time intervals $t \lesssim t_W$ the non-stationary kinetics of clusters of large enough size ($i \gg 1$) is described by the Zeldovich-Frenkel equation (Venables et al. 1984)

$$\partial_t g(i,t) = -\partial_i I(i), \quad I(i) = -D^i(i)[\partial_i g + g\partial_i F(i)], \tag{11.1.2}$$
$$g(i,0) = g_0(i),$$

where $g(i,t)$ is the local surface density of clusters with i adatoms, $D^i(i)$ is the diffusion coefficient in size space, $F(i)$ is the free energy of the formation of a cluster of size i. Different physical models of clusters and their growth (e.g. capillary, lattice, diffusion models etc.) lead to different approximations for $D^i(i)$ and $F(i)$.

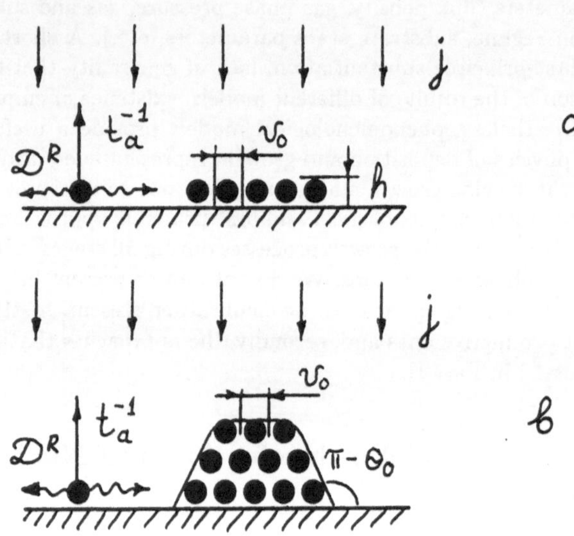

Fig. 10. Variants of cluster form for the capillary model of condensation: a) a disk-shaped cluster; b) a cupola-shaped liquid cluster with the contact angle θ_0.

For example, using the approximation of two-dimensional disk-shaped clusters (see Fig. 10) and LG model yields the following expression for $F(i)$ (Venables et al. 1984)

$$F(i)/k_BT = 2(b_d i)^{1/2} - i\ln(1+s) - \ln(n_c/n_1) + \frac{1}{2}\ln(2\pi i), \tag{11.1.3}$$

$$b_d = \Big(\frac{\lambda_0}{k_BT}\Big)^2 \frac{\pi v_0}{h}, \quad n = n_\infty \exp\left(-\frac{v_0}{h}\frac{q_3 - q_1 - q_2)}{k_BT}\right), \quad s = \frac{n_1}{n} - 1, \tag{11.1.4}$$

where q_1, q_2, q_3 are the specific free energies of the vapour-condensate, condensate-substrate, and substrate-vapor boundaries, respectively, λ_0 is the interphase free energy per unit length of liquid droplet boundary, v_0 is the volume occupied by one particle in a cluster, h is the height of a disk shaped cluster, n_c is the surface density of adsorption cells, n_∞ is the equilibrium density of the saturated vapor of adatoms. Since for flat clasters (layer-by-layer mode) one has $q_3 - q_1 - q_2 > 0$ the adatom density n corresponding to the beginning of the nucleation process is less than the equilibrium density n_∞. In (11.1.3) $2(b_d i)^{1/2}$ is the surface tension energy of a liquid droplet (we neglect the dependence of λ_0 on i), $\ln(s+1)$ is the difference between the chemical potentials of the gas of adatoms and the condensed phase at the substrate, $\ln(n_c/n_1)$ is the statistical Lothe-Pound correction induced by the specifics of distribution of n_1 adatoms over n_c places at the substrate, $\ln(2\pi i)/2$ is the correction due to the free energy change when separating a group of i molecules from the whole ensemble.

The free energy for cupolar-shaped nuclei has an analogous form (Venables et al. 1984)

$$F(i)/k_B T = 3[b_c^{\frac{1}{3}} i/2]^{2/3} - i\ln(1+s) - \ln(n_c/n_1) + \frac{1}{2}\ln(2\pi i), \qquad (11.1.5)$$

$$b_c = [v_0^2 (k_B T)^{-3}] \frac{4\pi}{3}(2+\cos\theta_0)(1-\cos\theta_0)^2 q_1^3, \qquad (11.1.6)$$

derived in the framework of a model of liquid droplets at the surface. The cupolar shape of nuclei is characteristic for the island mode of the film growth. Angle θ_0 entering (11.1.6) is the contact angle (see Fig. 10b). The Gibbs relation

$$q_3 = q_2 + q_1 \cos\theta_0 \qquad (11.1.7)$$

has been used to obtain (11.1.6).

It is important, that the function $F(i)$ has a typical form (see Fig. 11) with one maximum ($F(i_{cr})$) corresponding to the critical nucleus size i_{cr}. Thus, small clusters with $i < i_{cr}$ appear and disappear due to the thermodynamic fluctuations, while clusters with $i > i_{cr}$ grow regularly in average.

The following expression for the diffusion coefficient in dimension space is generally used

$$D^i(i) = L(i) D^R n_1 l_d^{-1}, \qquad (11.1.8)$$

with $L(i)$ being the length of a linear nucleus boundary, l_d - the length of a diffusional jump of an adatom, D^R - the coefficient of diffusion of adatoms in the physical space. A simple approximation of continuos ideal adsorbate gives for the latter quantity an expression

$$D^R = \frac{l_d^2 \nu}{\kappa} \exp\left(-\frac{\varepsilon_D}{k_B T}\right) \qquad (11.1.9)$$

with κ being the lattice coordination number - the number of nearest to an adatom adsorption places (for square lattice $\kappa = 4$). More realistic approximations for D^R have been discussed in Sect.10.2.

In the framework of approximations (11.1.3) and (11.1.5) the following expressions for characteristic parameters in the critical point have been derived (Venables et al. 1984)

$$i_{cr} = \frac{b_d}{\ln^2(s+1)}; \quad H(s) \equiv F(i_{cr}) = \frac{b_d}{\ln(s+1)} - \ln\frac{n_c}{n_1} + \frac{1}{2}\ln\frac{2\pi b_d}{\ln^2(s+1)}, \qquad (11.1.10)$$

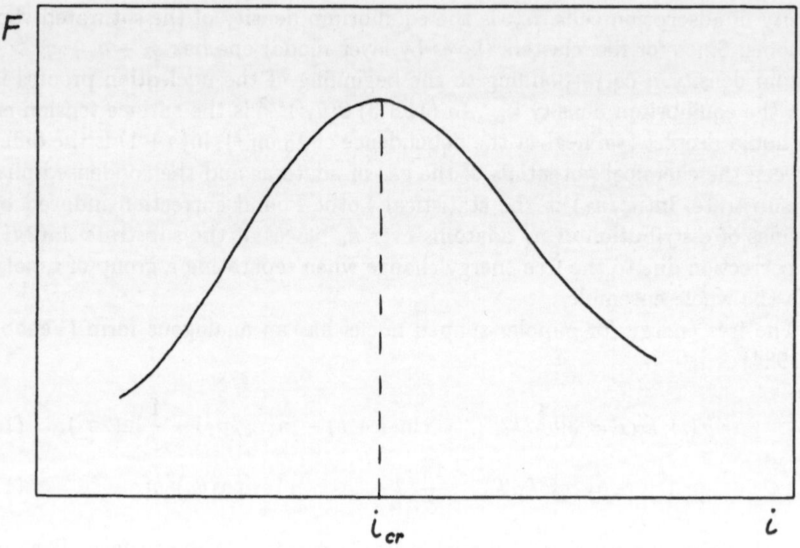

Fig. 11. A typical form of F versus i curve

$$i_{\rm cr} = \frac{2b_c}{\ln^3(s+1)}; \quad H(s) \equiv F(i_{\rm cr}) = \frac{b_c}{\ln^2(s+1)} - \ln\frac{n_c}{n_1} + \frac{1}{2}\ln\frac{4\pi b_c}{\ln^3(s+1)}. \qquad (11.1.11)$$

Here $H(s)$ has the sense of the activation barrier to a stable nucleus formation.

The quasistationary solution g^e to (11.1.2) with constant cluster flux $I = \text{const}$ and boundary condition $g^e \exp(F(i)/k_{\rm B}T) \to 0$ for $i \to \infty$ has the form

$$g^e(i) = I \exp\left(-\frac{F(i)}{k_{\rm B}T}\right) \int_i^\infty {\rm d}i' (D^i(i'))^{-1} \exp\left(\frac{F(i')}{k_{\rm B}T}\right). \qquad (11.1.12)$$

For $t \gg t_W$ and $i \ll i_{\rm cr}$ this distribution becomes equilibrium

$$g^e(i) = n_1 \exp\left(-\frac{F(i)}{k_{\rm B}T}\right), \quad i \to 0. \qquad (11.1.13)$$

Comparing (11.1.12) with (11.1.13) gives an expression for the quasistationary flux I

$$I = n_1 \left[\int_0^\infty {\rm d}i' (D^i(i'))^{-1} \exp\left(\frac{F(i')}{k_{\rm B}T}\right)\right]^{-1}. \qquad (11.1.14)$$

Expanding $F(i')$ in the vicinity of the critical point $i = i_{\rm cr}$ (point of maximum) yields an approximate reperesentation for I

$$I = n_1[-F''(i)/2\pi]^{\frac{1}{2}} D^i(i) \exp\left(-\frac{F(i)}{k_{\rm B}T}\right)\bigg|_{i=i_{\rm cr}}. \qquad (11.1.15)$$

11.2 Growth Stage

At times $t \gtrsim t_W$ the distribution function of undercritical clusters ($i < i_{cr}$) is described by the quasistationary distribution function (11.1.12) ($g = g^e$). For overcritical nuclei ($i > i_0 > i_{cr}$) one can neglect fluctuation processes of clusters decay and formulate the following initial-boundary value problem for $g(i,t)$ instead of (11.1.2)

$$\partial_t g(i,t) = -\partial_i[g(i,t)v(i,t)], \tag{11.2.1}$$

$$g(i,0) = g^0(i); \qquad g(i_0,t) = \left.\frac{I(s,i)}{v(s,i)}\right|_{i=i_0}, \tag{11.2.2}$$

where $v(s,i) = di/dt$ is the velocity of overcriticl nucleus growth, i_0 is the characteristic size to be found from the condition of the best separation of under- and overcritical nuclei. An approximation corresponding to some certain model of overcritical nuclei growth is to be used for evaluation of $I(s,i)$ and $v(s,i)$. Particularly, from (11.1.3), (11.1.5), and (11.1.15) one obtains the following representations for disk- and cupola-shaped nuclei:

$$I = C_d n n_c D^R (s+1) \ln^{3/2}(s+1) \exp\left[-\frac{b_d}{k_B T \ln(s+1)}\right], \tag{11.2.3}$$

$$I = C_c n n_c D^R (s+1) \ln^{5/2}(s+1) \exp\left[-\frac{b_c}{k_B T \ln^2(s+1)}\right], \tag{11.2.4}$$

with

$$C_d = (2v_0/\pi h l_d^2 b_d)^{\frac{1}{2}},$$

$$C_c = \sin\theta_0 \left[(3/2)^{\frac{1}{2}} \pi l_d^3 v_0^{-1}(1-\cos\theta_0)(2+\cos\theta_0)\right]^{-\frac{1}{3}} b_c^{-1}. \tag{11.2.5}$$

For $v(s,i)$ the following empirical representation, confirmed by a number of numerical calculations, is frequently used

$$v(s,i) = \Phi_1(s(t))\Phi_2(i). \tag{11.2.6}$$

For a lot of systems the following model of growth law is valid (Koropov and Sagalovich 1990)

$$\Phi_1(s) = s/t_a, \qquad \Phi_2(i) = A^{1/m} m i^{1-1/m}, \tag{11.2.7}$$

with $A \sim t_a^m$ being a dimensionless constant.

11.3 Coagulation Stage

For times $t \gtrsim t_C$ along with the process of isolated islands growth the process of islands coagulation begins. To describe the influence of this phenomenon on the film growth kinetics it has been proposed to include into the kinetic equation (11.2.1) a term representing the pair coagulation of liquid droplets (Fig. 12) in Smolukhovsky approximation (Osipov 1990, Voloshchuk 1984), introducing in such a way the liquid phase coverage $Z(t)$ of the surface ($Z(t)$ is the part of the surface area covered with the liquid phase). The corresponding initial - boundary value problem for the distribution function $g(i,t)$ can be put down in the form

$$\partial_t g(i,t) + \partial_i(vg) = \int_0^{i/2} \omega(i-i',i')g(i-i',t)g(i',t)\mathrm{d}i' - g(i,t)\int_0^{\infty} \omega(i,i')g(i',t)\mathrm{d}i', \qquad (11.3.1)$$

$$g(i_0,t) = I(s(t))[1-Z(t)]v^{-1}\big|_{i=i_0}, \qquad g(i,0) = 0, \qquad (11.3.2)$$

with ω being the kernel of coagulation operator.

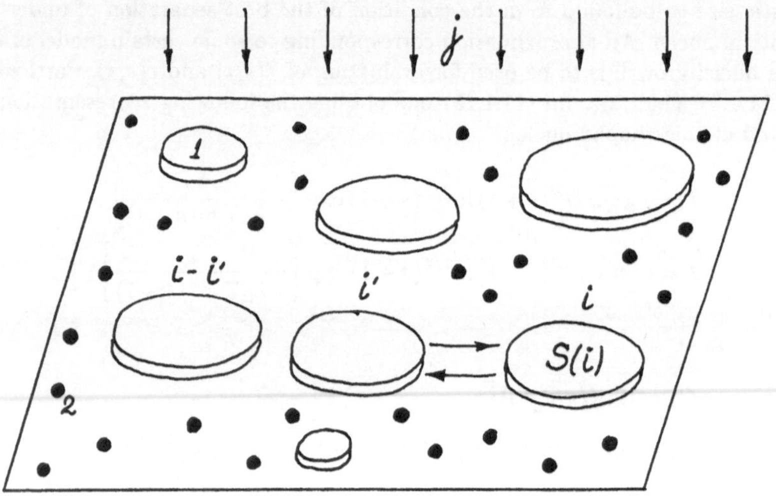

Fig. 12. Model of pair coagulation $(i-i')+i' \leftrightarrows i$ of liquid droplets at the surface: 2 – oversaturated vapor, 1 – condensate, $S(i) = \pi R^2(i)$.

Provided islands have the shape of an monoatomic disk or cupola, the surface coverage $Z(t)$ is coupled to $g(i,t)$ by the expression

$$Z(t) = \pi \int_{i_0}^{\infty} R^2(i)g(i,t)\mathrm{d}i, \qquad (11.3.3)$$

where $R(i)$ denotes the radius of the basis of a cluster of size i (the shape of the basis is assumed to be circular).

The problems (11.2.1), (11.2.2) or (11.3.1), (11.3.2) are to be completed with the material balance equation on the monomer density or on the supersaturation $s(t)$

$$s(t) = s(0) + \int_0^t \frac{s_0 - s(t')}{t_a}\mathrm{d}t' - \frac{1}{n}\int_{i_0}^{\infty} ig(i,t)\mathrm{d}i, \quad s_0 = \frac{jt_a}{n} - 1, \qquad (11.3.4)$$

where $s(0)$ and s_0 are the initial and the maximal supersaturations, respectively, j is the flux density of incoming particles ($j = \text{const}$).

The solution of (11.2.1), (11.2.2) or (11.3.1)–(11.3.3) together with (11.3.4) allows one to evaluate the main characteristics of a nucleating liquid film. Certain assumptions on mechanisms and process conditions gives one an opportunity to simplify significantly this evaluation. For example, under the approximation (11.2.6) the initial-boundary value problem is reduced to a closed kinetic equation on the supersaturation (Osipov 1990)

$$s(y) = s(0) + \int_0^y \frac{dx}{\Phi_1(s(x))}\left[\frac{s_0 - s(x)}{t_a} - \frac{i(y-x)I(s(x))}{n}\right], \qquad (11.3.5)$$
$$s(x) = s(t(x))$$

where $t(x)$ and $i(\rho)$ ($\rho = y - x$) are the inverse functions to the following ones

$$x(t) = \int_0^t \Phi_1(s(t'))dt', \qquad \rho(i) = \int_{i_0}^{i} \frac{di'}{\Phi_2(i')} \qquad (11.3.6)$$

The nucleation flux density $I(t)$, the total islands density $N(t)$, the monomer density $n_1(t)$, and the distribution function of nuclei sizes can be expressed through the solution to the integral equation (11.3.5):

$$I(t) = I(s(t)), \quad N(t) = \int_0^t I(t')dt', \quad n_1(t) = n(s(t)+1),$$
$$g(i,t) = I(s(x))[\Phi_1(s(x))\Phi_2(i)]^{-1} \quad (g(i,t) = 0,\ x \le 0), \qquad (11.3.7)$$
$$x(i,t) = y(t) - \rho(i).$$

In case of growing and pairwise coagulating droplets the material balance equation (11.3.4) can be presented as (Osipov 1990)

$$\dot{s}(t) = \left[j - \int_{i_0}^{\infty} v_0 g(i,t)di\right][n(1-Z)]^{-1} -$$
$$-(1+s)[t_a^{-1} - \dot{Z}(1-Z)^{-1}] - jZ(1-\alpha_0)[n(1-Z)]^{-1}, \qquad (11.3.8)$$

with α_0 being the probability that a monomer, attached to a cluster from the gas phase, increases the number of particles in this cluster by unity (for island growth $\alpha_0 = 1$, for layer-by-layer one – $0 \le \alpha_0 \le 1$). The system of equations (11.3.1)–(11.3.3) and (11.3.8) presents a closed kinetic model of the film growth incorporating the binary coagulation of droplets and effects of the increase of the surface coverage with liquid phase.

Assuming the clusters to be flat disks of the height h, relation (11.3.3) and material balance equation (11.3.8) can be rewritten in the form

$$Z(t) = (v_0/h)\int_{i_0}^{\infty} ig(i,t)di, \qquad (11.3.9)$$

$$\dot{s}(t) = \frac{j}{n} - \frac{s+1}{t_a} - \left[\frac{h}{v_0 n} - s - 1\right]\frac{\dot{Z}(t)}{1-Z}. \qquad (11.3.10)$$

From (11.3.1) one can readily derive a system of equations on the moments of the distribution function $g(i,t)$. Using a model growth law of islands

$$v = \frac{di}{dt} = is/t_0, \tag{11.3.11}$$

reduces the moment equations to one equation on $Z(t)$. The resulting closed system of equations on functions $Z(t)$ and $s(t)$ has the following form

$$\begin{cases} \dot{Z} = [s - G(s)]z + G(s) \\ \dot{s} = A - Bs - (C_T - s)(1 - Z)^{-1}[(s - G(s))Z + G(s)] \end{cases} \tag{11.3.12}$$

with

$$A = \left[\frac{jt_a}{n} - 1\right] B; \quad B = \frac{t_0}{t_a}; \quad C_T = \frac{h}{v_0 n} - 1; \quad G(s) = t_0 v_0 i_0(s) j(s)/n. \tag{11.3.13}$$

Within the model of one-dimensional substrate the problem (11.3.1), (11.3.2) has an analytical solution (Osipov 1990) provided

$$v = \hat{v}(t), \tag{11.3.14}$$

and the coagulation kernel may be represented as

$$\omega_{\mathrm{coag}}(R, R', t) = 2\hat{v}(t)\alpha(t)N_0/N(t), \tag{11.3.15}$$

with R and R' being the lenghts of one-dimensional islands, $\alpha(t)$ - a dimensionless function and N_0 - a fitting constant parameter with the concentration dimension.

12 Lateral Growth of Island Films

12.1 General Kinetic Model

Let us consider now an alternative kinetic scheme of two-dimensional islands coagulation (Dubrovskiy V. 1990), based on the geometric model by Kolmogorov (Belen'kiy 1980). It is applicable for crystalline islands and describes simultaneous coagulation of a few islands, provided the form of growing islands remains unchanged (see Fig. 13) (the model of cristalline coagulation). We shall formulate the kinetic model for crystalline coagulation assuming that cluster lateral growth is defined entirely by the cluster perimeter.

Let S and $S_f(t)$ be the total surface area and the area free from the adsorbate. Introducing the adatom density $n_1(t)$ on free area and the fractional coverage g of the free area yields

$$g(t) = \frac{S_f(t)}{S}, \quad z(t) = 1 - g(t), \tag{12.1.1}$$

$$n_1(t) = \frac{N_1(t)}{S_f(t)} = \frac{N_1(t)}{Sg(t)}, \tag{12.1.2}$$

where $N_1(t)$ is the total number of free adatoms at the surface, $z(t)$ is the fractional coverage of monolayer film. The nucleation starts when the adatom density exceeds the equilibrium one $n_2^0 = \theta_2^0/\sigma$ and phase 2 becomes metastable. For low enough temperatures ($\theta_2^0 \ll 1$) the formula (9.4.20) is valid and the oversaturated vapor of adatoms is ideal, the coverage of the substrate with phase 1 is close to zero, the surface energies $q(\theta_2^0)$ and $q(\theta_1^0)$ are approximately equal to the energies of

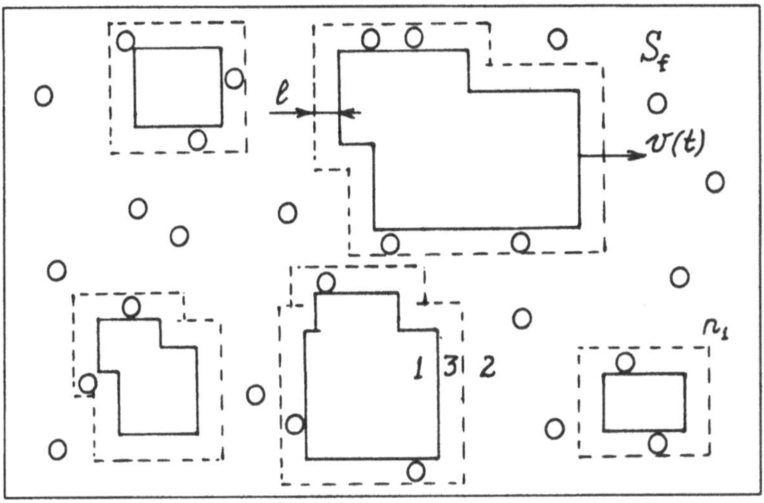

Fig. 13. Model of solid phase island coagulation at final stage of continuous film formation. 2 – free surface area, 1 – filled area, 3 – "feeding layer" of width l.

phases separation for substrate-gas q_{sg}, and substrate-condensate and condensate-gas (q_{sc}, q_{cg}), respectively. For the sake of breavity we shall denote $n \equiv n_2^0$ and $\theta \equiv \theta_2^0$ assuming that indexless quantities refer to the saturated two-dimensional vapor.

Using definitions (12.1.1) and (12.1.2) and assuming that at $t = 0$ $n_1(0) > n$, $g(0) = 1$ one can present the material balance equation in monolayer in the form

$$n_1(0) + \int_0^t dt' g(t')(j - n_1(t')/t_a) = n_1(t)g(t) + z(t)/\sigma. \tag{12.1.3}$$

The differential form of this equation can be obtained using the condition $\sigma n(t) \ll 1$:

$$\partial_t n_1(t) = j - n_1(t)/t_a + \sigma^{-1}\partial_t \ln g(t). \tag{12.1.4}$$

The Kolmogorov model of crystalline growth couples $g(t)$ with the nucleation rate on a free part of the substrate $\alpha(t)$ ($\alpha(t) = I(t)$, $i_0 \to 0$) and velocity of linear growth of the crystalline phase v (Belen'kiy 1980):

$$g(t) = \exp(-F(t)), \tag{12.1.5}$$

$$F(t) = C \int_0^t dt' \alpha(t')\rho^2(t', t), \quad \rho(t', t) = \int_{t'}^t dt'' v(t''), \tag{12.1.6}$$

with C being the geometrical factor, dependent on the form of an isolated island. The last two formulae lead to the equations

$$v^{-1}(t)\partial_t F(t) = 2C \int_0^t dt' I(t')\rho(t', t) \equiv l(t), \tag{12.1.7}$$

$$[2Cv(t)]^{-1}\partial_t l(t) = \int_0^t dt' I(t') = N(t), \tag{12.1.8}$$

with $N(t)$ being the total number of clusters at moment t, $l(t)$ – the crystallite perimeter related to the unit area of the free surface, so that

$$L(t) = l(t)g(t) \tag{12.1.9}$$

has the sense of the crystallisation front perimeter of two-dimensional film.

For $I(t)$ we shall use the expression (11.2.3). The growth velocity $v(t)$ can be evaluated in the following way. The coverage is changed via processes of evaporation and absorption. The corresponding rate of the coverage change is

$$-\partial_t g = \frac{\sigma l_d}{\zeta t_d}(n_1(t) - n)L(t), \tag{12.1.10}$$

with

$$t_d = \nu^{-1} \exp\left(\frac{E_d}{k_B T}\right), \tag{12.1.11}$$

where t_d presents the mean time interval between two succesive diffusion "jumps" with the mean length t_d. Combining (12.1.5), (12.1.7), and (12.1.9) gives the following expression for the velocity of growth

$$v(t) = rD^R ns(t), \tag{12.1.12}$$

where r is the mean radius of cluster (Belen'kiy 1980). One can introduce more complicated relations, e.g.,

$$v(t) = v_0 U(s(t)), \tag{12.1.13}$$

where U is a function, depending on the growth mechanism involved.

One can apply the Kolmogorov model provided a growing film has crystalline structure. It is known (Dash 1975) that mechanism gas-crystall is involved in a crystalline film formation at temperatures $T \lesssim 0.3T_m$ with T_m being the melting temperature of the deposit. This mechanism takes place almost always in technological processes of film deposition onto "cold" substrate (Dash 1975, Jaycock and Parfitt 1981). If $0.3\ T_m \lesssim T \lesssim 0.6\ T_m$, the intermediate regime occurs, while for higher temperatures ($T \gtrsim T_m$) film formation proceeds via the scheme: gas $\rightarrow$ liquid $\rightarrow$ crystall. The corresponding model of liquid droplets coagulation has been considered in the previous Chap.

A model of nucleation kinetics together with the Kolmogorov model of layer growth allows one to evaluate the function $g(n_1)$. Inserting (1.2.3), (12.1.2) into (12.1.24)-(12.1.28) yields main parameters of the thin film growth in terms of the supersaturation s:

$$\theta^{-1}F(x) = s(0) - s(x) + \int_0^x dx'(s_0 - s(x')), \tag{12.1.14}$$

$$l(x) = \frac{Cr}{D^R t_a}[s_0/s(x) - 1 - s'(x)/s(x)], \tag{12.1.15}$$

$$N(x) = (2D^{R^2} t_a^2 n)^{-1} l'(x)/s(x), \tag{12.1.16}$$

$$\theta = \sigma n, \quad x = t/t_a.$$

The material balance equation (12.1.3) together with (11.2.3) and (12.1.12) leads to the following closed nonlinear equation on the supersaturation

$$\frac{\mathrm{d}s(x)}{\mathrm{d}x} = s_0 - s(x)[1 + 4\kappa^3 \int_0^x \mathrm{d}x'\psi(s(x')) \int_{x'}^x \mathrm{d}x'' s(x'')], \tag{12.1.17}$$

$$\psi(s) = I(x)(2D^R n^2)^{-1}, \quad D^R = \frac{Crl_0}{\zeta t_d} \approx \frac{l_0^2}{t_d}, \quad \kappa = D^R t_a n = l_0^2 n,$$

where D^R is the classical diffusion coefficient, l_0 is the length of a single diffusional jump, l_d is the mean diffusion length. Equation (12.1.17), describing the growth with coagulation of any number of clusters, is identical to (11.3.5), where growth law is to be taken from (12.1.12) that does not include coagulation effects. Note, that $F(x)$ has the sense of effective coverage, i.e., the arythmetic sum of the coverages of growing isolated crystallites. When $F(x) \ll 1$ there is no coagulation and our function is identical to true coverage. Thus, the kinetics of the Kolmogorov model at the first stages of film growth is the same as that described by the kinetic equation for overcritical nuclei with boundary condition at $i = i_0$. It can be shown that appropriate physical models of nucleation within the Kolmogorov model lead to the corresponding kinetics of initial stage, that can be formally obtained by a substitution $\exp(-F(x)) \approx 1 - F(x)$.

12.2 Kinetic Model of Multilayer Film Growth

Let us describe a phenomenological model of the multilayer film growth that is a simplified version of the general kinetic model of multilayer adsorption presented in Chap.6. The latter model allows one to evaluate the distribution function $\theta(\beta, \boldsymbol{R}, t)$ by solving general equation (6.1.19) and then to calculate main parameters of a multilayer adsorption film (h, Δ, Ω).

Suppose that the covearge of the first layer $\theta(1,t)$ can be expressed through $F_1(x)$ by

$$\theta(1,t) = 1 - \exp(-F_1(t)), \tag{12.2.1}$$

$$F_1(0) = 0, \quad F_1(t) \underset{t\to\infty}{\to} \infty. \tag{12.2.2}$$

The function $F_1(t)$ describes the kinetics of the first layer formation and can be defined by solving the Kolmogorov model equation (12.1.14). We shall use here more simple approximations for time dependence of $F_1(t)$, namely (Belen'kiy 1980, Kashchiev 1977)

$$F_1(t) = (\omega_1 t)^{m_1}, \tag{12.2.3}$$

with $m_1 \geq 0$ and $t_1 = 1/\omega_1$ being the characteristic time of filling the first monolayer. Note that the solution of (12.1.4) gives analogous asymptotic behavior for $F_1(t)$:

$$F_1(t) \to (j - \frac{n}{t_a})\sigma t, \quad x \gg x_1. \tag{12.2.4}$$

Following the ideas by Wetter (1967), let us postulate a system of kinetic equations on the coverages of succesive layers

$$\theta(\beta+1) = \int_0^t \mathrm{d}t' G_{\beta+1}(t-t')\theta(\beta,t'), \tag{12.2.5}$$

$$G_{\beta+1}(t) = -\frac{\mathrm{d}}{\mathrm{d}t}\exp(-F_{\beta+1}(t)). \tag{12.2.6}$$

The physical sense of this model is that the coverage of the layer number $\beta+1$ depends only on its own kinetic parameters ($\omega_{\beta+1}$, $F_{\beta+1}$) and on the coverage of the preceeding layer. Introducing coverage differencies $\tilde{\theta}(\beta,t) = \theta(\beta,t) - \theta(\beta+1,t)$, and performing the Laplace transforms of $\theta(\beta,t)$ and $G_\beta(t)$

$$\theta(\beta,\omega) = \int_0^\infty \mathrm{d}t \exp(-\omega t)\theta(\beta,t);$$

$$G(\beta,\omega) = \omega\int_0^\infty \mathrm{d}t \exp(-\omega t - F_\beta(t)), \tag{12.2.7}$$

gives the following expression for $\tilde{\theta}(\beta,\omega)$

$$\tilde{\theta}(\beta,\omega) = G(\beta+1,\omega)\prod_{\beta'=2}^{\beta}[1-G(\beta',\omega)]\tilde{\theta}(1,\omega)$$

$$= \frac{G(\beta+1,\omega)}{\omega}\prod_{\beta'=2}^{\beta}[1-G(\beta',\omega)]. \tag{12.2.8}$$

Thus, provided the functions $F(\beta,t)$ are known, one can perform the inverse Laplace transform and obtain functions $\tilde{\theta}(\beta,t)$

$$\tilde{\theta}(\beta,t) = (2\pi i)^{-1}\int_{-i\infty}^{+i\infty} \mathrm{d}\omega \exp(\omega t)\tilde{\theta}(\beta,\omega). \tag{12.2.9}$$

After that all the characterictics of a multilayer film can be evaluated. For example, when $F_1(t) = F_2(t) = \ldots = \omega_1 t$, $h_1 = h_2 = \ldots = h$ (normal growth law (Kashchiev 1977)) one can readily perform the Laplace transforms and get

$$\tilde{\theta}(\beta,\omega) = \frac{G(\beta+1,\omega)\omega_1^\beta}{\omega(\omega+\omega_1)^{\beta+1}},\quad \tilde{\theta}(\beta,t) = \frac{(\omega_1 t)^\beta}{\beta!}\exp(-\omega_1 t), \tag{12.2.10}$$

$$\bar{h}(t) = h\omega_1 t,\quad \Delta^2 = h^2\omega_1 t. \tag{12.2.11}$$

This means that for the normal growth law the relief distribution is the Poisson one, the mean height of the film grows linearly with time, while mean variation growth law is $\Delta \sim t^{1/2}$. The ratio of the squared variation to the mean height is constant during the growth

$$\zeta(t) \equiv \left(\frac{\Delta^2(t)}{h^2}\right)\Big/\left(\frac{\bar{h}(t)}{h}\right) = 1. \tag{12.2.12}$$

In more realistic approximation based on some ideas of BET-model of adsorption isotherms (Flood 1967) it is assumed that $h_1 \neq h_2 = h_3 = \ldots$, $F_1 \neq F_2 = F_3 = \ldots$, $\omega_1 \neq \omega_2 = \omega_3 = \ldots$ and function $F_1(t)$ is an arbitrary one. In this situation the following expressions for film parameters can be obtained (Dubrovskiy V. 1990)

$$\bar{h}(t)/h_2 = \nu\theta(1,\omega_1 t) - A_{12}(\omega_1 t) + \omega_2 t, \tag{12.2.13}$$

$$\Delta^2(t)/h_2^2 = \nu\left[\nu\theta(1,\omega_1 t) - 2A_{12}(\omega_1 t) + 2\omega_2 t\right]\exp(-F_1(\omega_1 t))$$
$$+A_{12}(\omega_1 t)\left[A_{12}(\omega_1 t)(2B_1(\omega_1 t) - 1) - 1\right] + \omega_2 t, \qquad (12.2.14)$$

$$\theta(1,\omega_1 t) = 1 - \exp(-F_1(\omega_1 t)), \quad \nu = h_1/h_2. \qquad (12.2.15)$$

Functions $A_{12}(x)$ and $B_\beta(x)$ in (12.2.13)-(12.2.15) are defined by the following expressions

$$A_{12}(x) = \frac{\omega_1}{\omega_2}K_1(x), \quad B_\beta(x) = I_\beta(x)/K_\beta^2(x), \qquad (12.2.16)$$

$$K_\beta(x) = \int_0^x dx' \exp(-F_\beta(x')), \quad I_\beta(x) = \int_0^x dx' x' \exp(-F_\beta(x')). \qquad (12.2.17)$$

When $\omega_1 t \gg 1$ one can simplify the expressions above to obtain

$$\frac{\bar{h}(t)}{h_2} = \nu - A_{12}(\infty) + \omega_2 t, \quad \frac{\Delta^2}{h^2} = A_{12}(\infty)[A_{12}(\infty)(2B_1(\infty)-1)-1] + \omega_2 t. \qquad (12.2.18)$$

Thus, asymptotic behaviour (12.2.18) of $\bar{h}(t)$ and $\Delta^2(t)$ differs from (12.2.13) only by the shift. Parameter $\zeta(t)$, calculated using (12.2.18), also tends to the unity when $\omega_1 t \gg 1$. This proves that asymptotic relief within this model does not depend on the first layers kinetics and is goverened by the normal growth law of the top layers.

The asymptotic behavior of functions $\bar{h}(t)$ and $\Delta^2(t)$ for $\omega_2 t \gg 1$ may be obtained for an arbitrary function $F_2(\omega_2 t)$ as well

$$\frac{\bar{h}(t)}{h_2} = (\nu + B_2(\infty) - 1)\tilde{\theta}(1,\omega_1 t) - C_{12}(\omega_1 t) + \frac{\omega_2 t}{K_2(\infty)}, \qquad (12.2.19)$$

$$\frac{\Delta^2(t)}{h_2^2} = B_2(\infty)(B_2(\infty) - 1)\tilde{\theta}(1,\omega_1 t) + (\nu + B_2(\infty) - 1)\exp(-F_1(\omega_1 t))$$
$$\times\left[(\nu + B_2(\infty) - 1)\tilde{\theta}(1,\omega_1 t) - 2C_{12}(\omega_1 t) + 2C_{12}(\omega_1 t) + \frac{2\omega_2 t}{K_2(\infty)}\right]$$
$$+ C_{12}(\omega_1 t)[C_{12}(\omega_1 t)(2B_1(\omega_1 t) - 1)$$
$$- (2B_2(\infty) - 1)] + (2B_2(\infty) - 1)\frac{\omega_2 t}{K_2(\infty)}, \qquad (12.2.20)$$

$$C_{12}(x) = \frac{A_{12}(x)}{K_2(\infty)} = \frac{\omega_2 K_1(x)}{\omega_1 K_2(\infty)}. \qquad (12.2.21)$$

Investigation of the expressions (12.2.19), (12.2.20) allows one to conclude that they have correct behaviour at $t = 0$ ($\bar{h}(0) = \Delta(0) = 0$), are identical to (12.2.13), (12.2.14) when $F_2 = \omega_2 t$ and become identical to (12.2.11) when $F_1 = \omega_1 t$, $\omega_1 = \omega_2$ and $\nu = 1$. For $\omega_2 t \gg 1$

$$\zeta(t) \to q(\infty) = 2\int_0^\infty dx\, x \exp(-F_2(x))$$
$$\times\left\{\left[\int_0^\infty dx \exp(-F_2(x))\right]^2 - 1\right\}^{-1} \neq 1, \qquad (12.2.22)$$

i.e., the quantity $q(\infty)$ depends on the velocity of upper layers filling.

Using power approximations $F_\beta(x) = x^{m_\beta}$ $(\beta = 1,2)$ yields

$$q(\infty) = \Gamma(1+2/m_2)/\Gamma^2(1+1/m_2) - 1, \tag{12.2.23}$$

with $\Gamma(y)$ being the gamma-function.

From (12.2.23) it is clear that $q(\infty)$ decreases with the increase of m_2. The power law for lateral growth velocity $v(x) = x^{\mu}$ gives the relation

$$m_2 = 2\mu_2 + 2. \tag{12.2.24}$$

The diffusional growth law ($\mu_2 = -1/2$) leads to $m_2 = 1$ and $q(\infty) = 1$. When $\mu_2 < -1/2$, $q(\infty) > 1$ and relief becomes more inhomogenious. When $\mu_2 > -1/2$, $q(\infty) < 1$ and the film becomes more flat.

13 Calculations of Thin Films Growth Kinetics

13.1 Capillary Model

Let us discuss the results of calculations made on the base of the capillary model, presented in Chap.12, for disk-shaped nuclei growing in diffusion regime. Under these assumptions the following approximations for functions $\Phi_1(s)$, $\Phi_2(i)$ in (11.2.6) are valid

$$\Phi_1(s) = \frac{s}{t_a}, \quad \Phi_2(i) = 2\pi D^R t_a n z(i) K_1(z(i))/K_0(z(i)), \tag{13.1.1}$$
$$z(i) = (i\Omega/\pi h D^R t_a)^{\frac{1}{2}},$$

with $K_i(x)$ being the McDonald function, Ω – the volume of cluster of height h.

Function $i(\rho)$ is the inverse one to $\rho(i)$ defined as

$$\rho(i) = (h/\Omega n)\int_0^{z(i)} dx K_0(x)/K_1(x). \tag{13.1.2}$$

Solving numerically (11.3.5) with (11.2.3) gives function $s(y)$. Then, using (11.3.7), we obtain all the characteristics of condensation process. The following values of constants have been used for the calculations presented in Figs. 14–17:

$$(D^R tu)^{\frac{1}{2}} = 10^{-8}\mathrm{m}, \quad b_d = 40, \quad l_d = 2\times 10^{-10}\mathrm{m},$$
$$n_0 = l_d^{-2}, \quad s_0 = 10, \quad n = 10^{17}\mathrm{m}^{-2}, \quad hl_d^2/\Omega = 1,$$

where $u = R/R_{cr}$ with R_{cr} being a critical radius (see below). The behavior of $s(t)$ at large t (Fig. 14) fits qualitatively results by Zinsmeister (Zinsmeister 1968, 1969, 1971). When $\Phi_2(i) \sim i^{1/2}$ (i.e., when $m = 2$ in (11.2.7)), the total number of stable nuclei $N \to$ const for $t \to \infty$ (Fig. 17), in contrast to the results by Zinsmeister (Zinsmeister 1968, 1969, 1971), where $N \approx \ln t$. Newertheless, experiments show that for large t N does tend to a constant. That confirms the validity and accuracy of the presented model. The distribution function q has the shape qualitatively coinsiding with the experimental curves (Lewis and Anderson 1978, Kern et al. 1987, Bauer at al. 1966), while the method of Zinsmeister (1968, 1969, 1971) gives a worse agreement. In accord to the experiments (Zinsmeister 1968, 1969, 1971) our model leads to a weak dependence of N versus T.

In the case of an impulse source the main equation on the supersaturation can be presented in a form

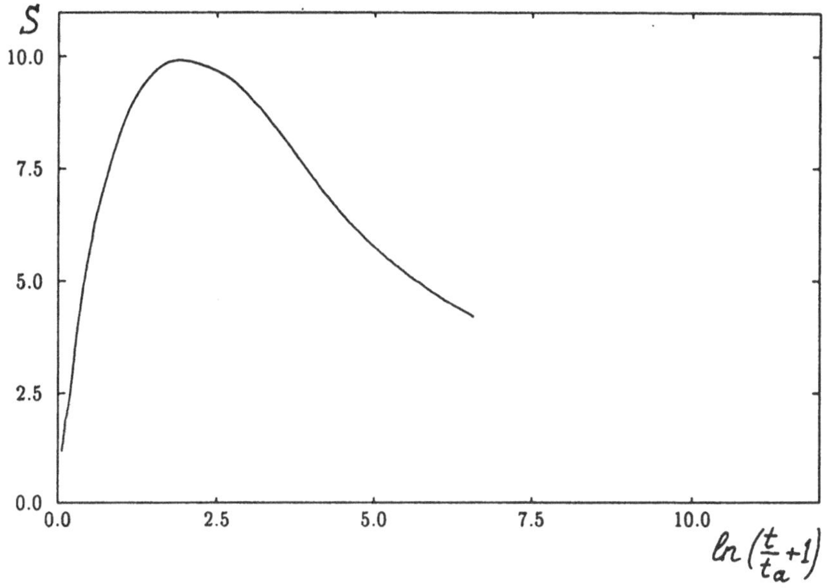

Fig. 14. Time dependence of the supersaturation.

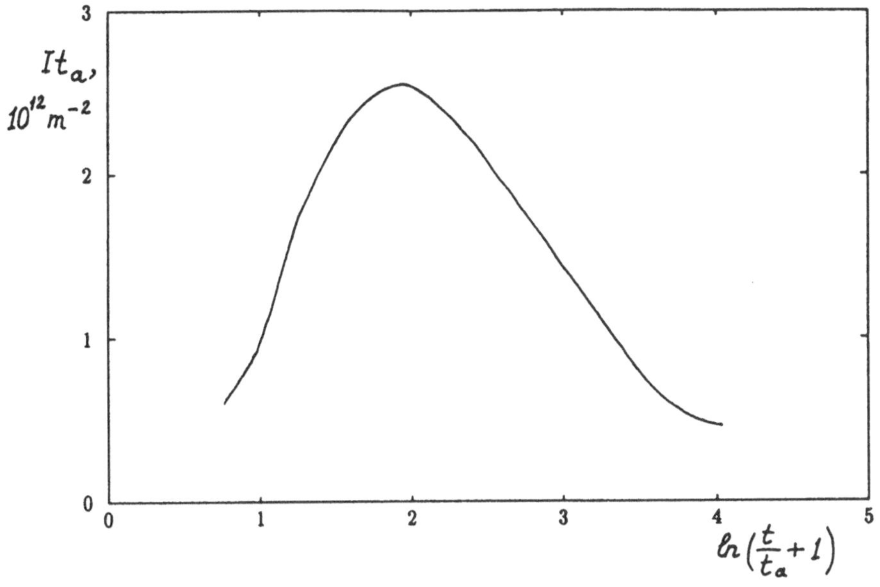

Fig. 15. Time dependence of the nucleation rate.

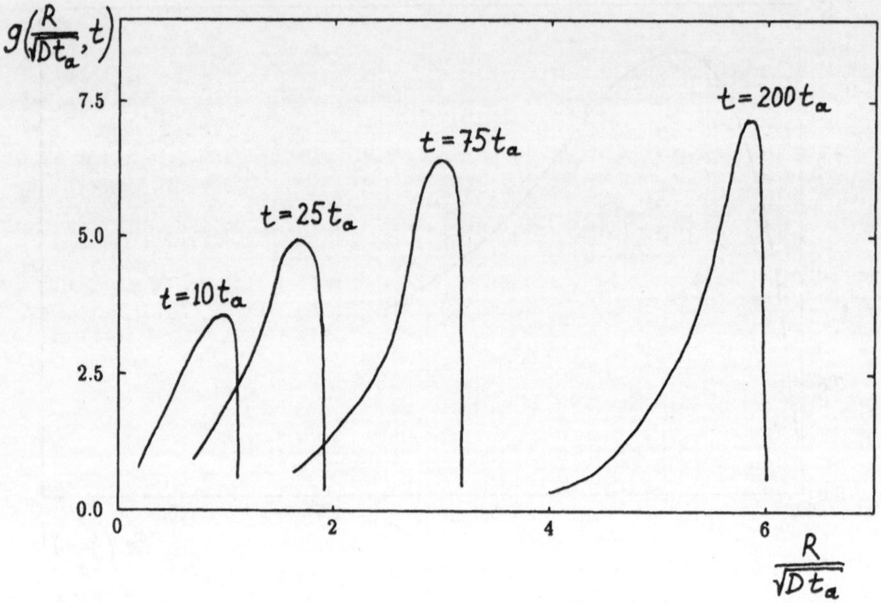

Fig. 16. Distribution function of nuclei over radii at different time moments.

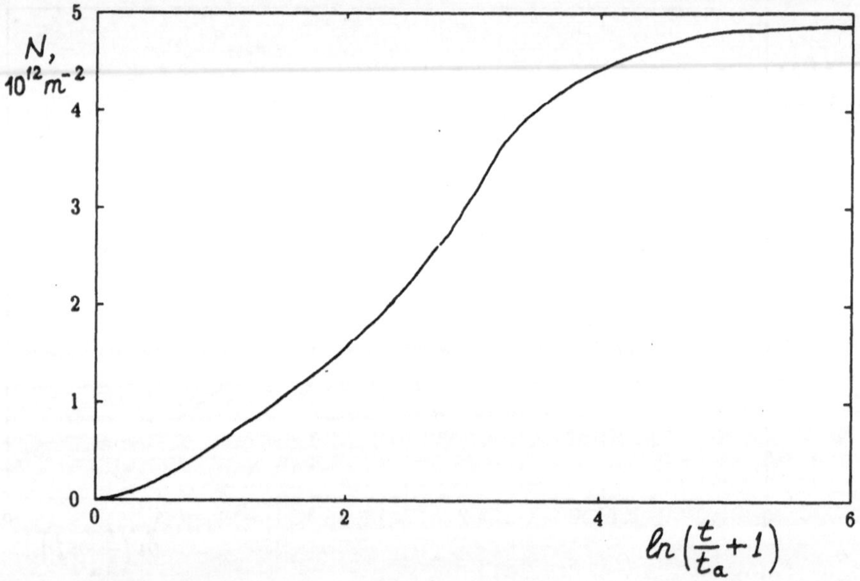

Fig. 17. Total number of stable nuclei per unit substrate area as a function of time.

$$s(y) = s(0) - \int_0^y dx \Phi_1^{-1}(s(x)) \left[\frac{s(x)+1}{t_a} + \frac{i(y-x)I(s(x))}{n} \right]. \tag{13.1.3}$$

By analogy with (11.2.7) the following model for clusters growth law will be assumed

$$\Phi_1(s) = \frac{s}{t_0}, \quad \Phi_2(i) = A^{1/m} m i^{m-1/m}, \tag{13.1.4}$$

with t_0 being some characteristic time scale, $A \sim t_0^m$ - a dimensionless constant. Let us introduce quantities $\mu = m/d$, $u = R/R_{cr}$, $R = Ci^{1/d}$, $R_{cr} = Cb_d^{1/d} s^{-1}$ ($d = 2, 3$), $\tau = (\mu + 1)\ln[R_{cr}(t)/R_{cr}(t_2)]$, and normalized to the unity distribution function of clusters over dimensions $G_\mu(u)$ ($\int_0^{u_0} G_\mu(u)du = 1$). Figure 18. illustrates the comparison of distribution function $G_1(u)$ (with $d = 3$) obtained from approximate solution of (13.1.3), with the experimental one (Hermann and Rhodin 1966) (for the systems Au-SiO_2, Au-NaCl, and Ni-KCl). The comparison of experimental data on Sn-C (Blackman and Guzzon 1959) and Au-NaCl (Hermann and Rhodin 1966) with theoretical curve ($d = 3$, $\mu = 2$) is presented in Fig. 19 and shows a good agreement.

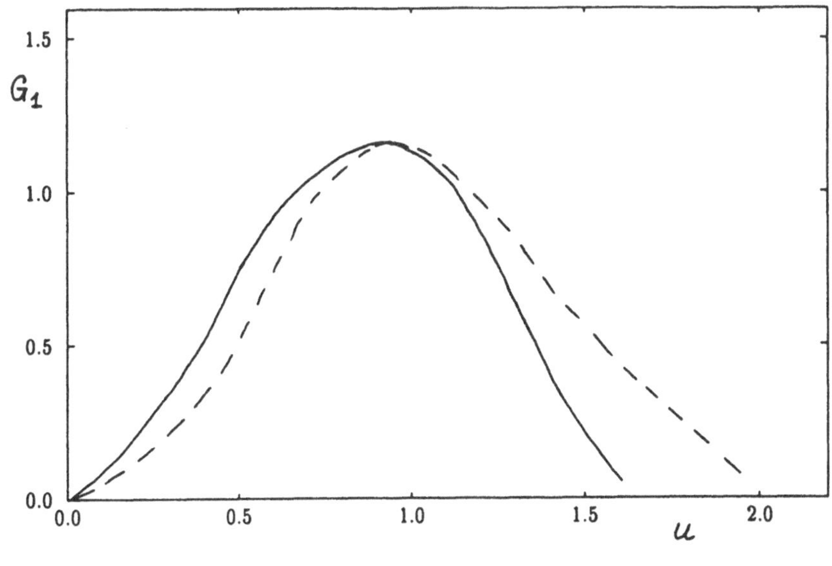

Fig. 18. Comparison of theoretical and experimental results on nuclei distribution. Solid line presents theoretical dependence of $G_1(u)$ with $d = 3$, dashed - experimental one (Hermann and Rhodin 1966).

In a number of experiments substrate temperature is kept to be so low, that mean life time t_a of adatoms exceeds the time of maximum supersaturation formation (so-called regime of complete condensation). In this case one is to solve the condensation equation together with the non-stationary diffusion equation describing the growth of clusters. Taking $t_a = \infty$ gives the following balance equation

$$s(y) = \int_0^y \frac{j}{n\Phi_1(s)} dx - \int_0^y \frac{i(y-x)I(s(x))}{n\Phi_1(s(x))} dx. \tag{13.1.5}$$

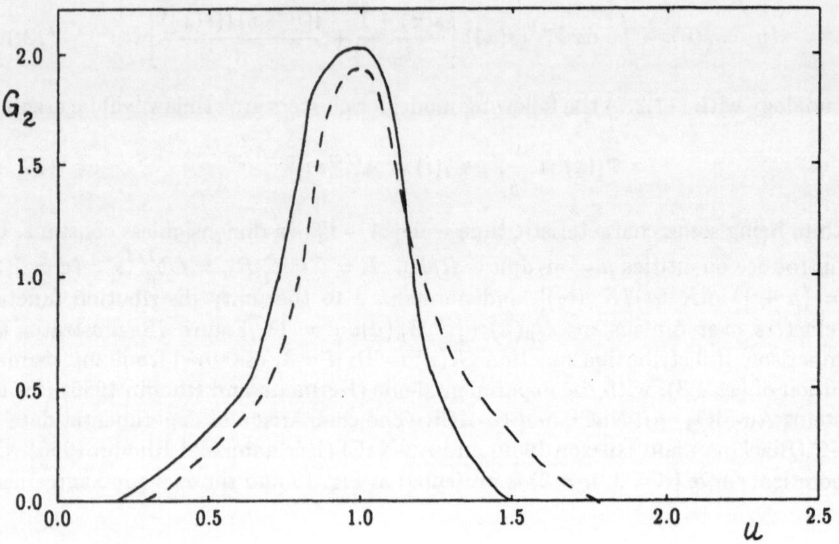

Fig. 19. Comparison of theoretical and experimental results on nuclei distribution. Solid line presents theoretical dependence of $G_2(u)$ with $d = 3$, dashed – experimental one (Hermann and Rhodin 1966).

For disk-shaped islands one has

$$i(y-x) = 4\pi D t_0 h(y-x)\Omega^{-1}, \qquad \Phi_1(s) = \mu^2(s)/t_0. \tag{13.1.6}$$

Figs. 20–22 present the results of numerical solution of (13.1.5), (13.1.6) under the following conditions:

$$n = 10^{17}\mathrm{m}^{-2},\ \ D^R = 10^{-19}\mathrm{m}^2/\mathrm{s},\ \ h/\Omega = 2.5 \times 10^{19}\mathrm{m}^{-2},$$
$$b_d = 40,\ \ l_d = 2 \times 10^{-10}\mathrm{m},\ \ n_0 = l_d^{-2}$$

for three different incoming flows: (1) $j = 0.1 \times j_0$, (2) $j = j_0$, (3) $j = 10 \times j_0$, with $j_0 = 10^{16}\mathrm{m}^{-2}/\mathrm{s}$. The main features of the condensation in this regime are: the existence of a sharp maximum of the supersaturation and the non-zero value of quantity $s(\infty)$, due to the condition $\Phi_2(i) = \mathrm{const}$.

To take into account the coagulation processes equation (11.3.1) may be used, while (11.3.8) may be approximated by

$$\dot{s} = \frac{s_0 - s}{t_a} - \left[\int_0^\infty u(R) v_R g(R,t)\mathrm{d}R - \frac{dz(1+s_0)}{t_a}\right](1-z)^{-1}, \quad d = (2,3), \tag{13.1.7}$$

due to the fact that the adatom concentration is generally much less than the surface concentration of particles at cluster foot. In (13.1.7) $s_0 = jt_a/n - 1$, $u(R) = (\Omega n)^{-1}\mathrm{d}V/\mathrm{d}R$, with $V(R)$ being the volume of a cluster with the foot radius R, $v_R = \dot{R}(t)$ (for disk-shaped clusters $u(R) = 2\pi Rh(\Omega n)^{-1}$, for cupola-shaped ones - $u(R) = (\pi/2)(R^2/\Omega n)\cos^{-3}(\theta_0/2)\sin(\theta_0/2)(2+\cos\theta_0)$).

It is convenient to express coalescence kernel in terms of R, t variables

$$\omega(R,R') = P(R+R',t)(v_R + v_{R'})N^{-1}(t) + \beta_0(R^{-q} + R'^{-q}), \tag{13.1.8}$$

with the first term describing the coalescence via the lateral growth of clusters, while the second one - via clusters lateral diffusion (Kashchiev 1976), $P(R,t)\mathrm{d}R$ being the probability to find a nearest cluster at the distance in the range $(R, R+\mathrm{d}R$ from some fixed cluster (we consider only nearest neighbour binary coagulation), β_0 and q being the model parameters.

In a simple case of randomly distributed islands one has

$$P(R,t) = 2\pi RN(t)\exp(-\pi R^2(t)N(t)). \tag{13.1.9}$$

Thus, the coagulation kernel of immovable, randomly distributed over the substrate clusters is an essentially non-isotropic function

$$\omega(R,R') = 2\pi(R+R')(v_R + v_{R'})\exp[\pi(R+R')^2 N(t)]. \tag{13.1.10}$$

Even more complicated expression can be obtained, provided the correlations in substrate distribution of clusters are taken into account (Trofimov 1975).

Equations (13.1.23) and (13.1.7) with an eye to (11.2.3), (13.1.1), and (13.1.10) have been solved numerically with the following parameters:

$$s(0) = R_0 = \alpha_0 = \beta_0 = 0, \quad h/\Omega n = 4 \times 10^{-3},$$
$$s_0 = 10, \quad t_a = 10^{-3}\mathrm{s}, \quad D = 10^{-15}\mathrm{m}^2/\mathrm{s},$$

with the results presented in Fig. 23. It is seen, that there is a possibility of a second maximum formation in accordance with some experiments (Lewis and Anderson 1978, Morris and Hines 1970, Sacedon and Martin 1972). Figs. 24 and 25 present the behavior of $z(t)$ and $s(t)$ for more special model (Osipov 1990). Fig. 26 illustrates a typical time dependence of the nucleation rate. The effect of secondary nucleation is also demonstrated.

13.2 Generalized Kolmogorov Model

Let us discuss the results of numerical solution of the main equation (12.1.17) of the Kolmogorov model (see Sect.12.1) for the monolayer film growth, presented in Figs. 27–32. The crystallization front parameters and total number of islands are found from (12.1.15), (12.1.16). Parameter $x_0 \gtrsim 1$ determines time scale of the maximum supersaturation s_0 formation in the absence of nucleation, x_1 is a characteristic time of the nucleation stage. In incomplete nucleation regime $x_0 \ll x_1$, the maximum of the supersaturation $s(x_m)$ is close to the "theoretical maximum" s_0. In complete condensation regime $x_0 > x_1$, i.e. nucleation finishes by the time t_a, the maximum of the supersaturation is much less than s_0. After the nucleation is over ($x \gtrsim 2x_1$), the total number of islands becomes constant. This result meets the conclusions made on the base of the capillary model (see above) and differs from the results of (Zinsmeister 1971, Venables 1973), where $n_1(x)$ was errorneously assumed to be a constant, that led to $N(x) \to \infty$. It should be emphasized here that $N(x)$ is a total number of cluster centers, nucleated at the surface prior to the coagulation process beginning, rather than the real number of isolated clusters

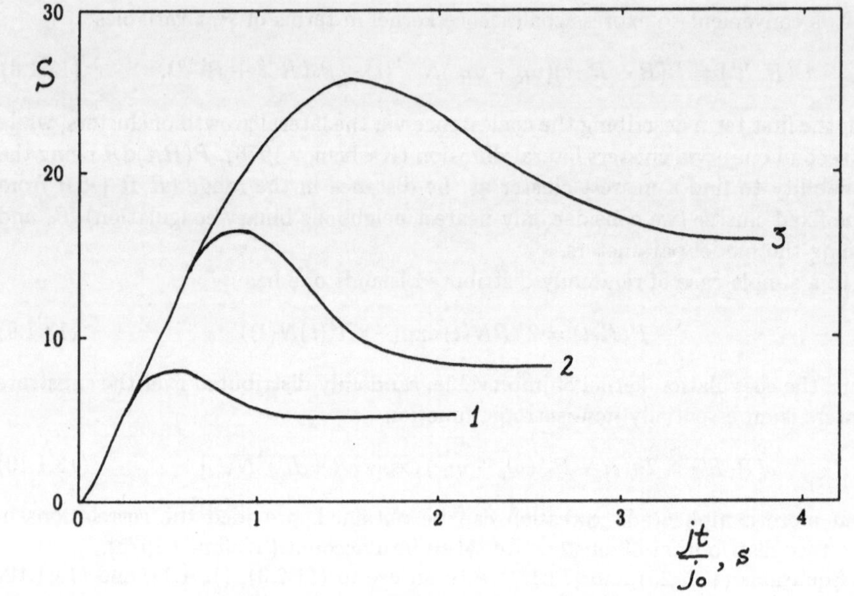

Fig. 20. Time dependence of the supersaturation for nonstationary diffusion growth of two-dimensional nuclei for the three different intensities of adatom source: 1 - $j = 0.1 j_0$, 2 - $j = j_0$, 3 - $j = 10 j_0$ with $j + 0 = 10^{16} \mathrm{m}^{-2}\mathrm{s}^{-1}$

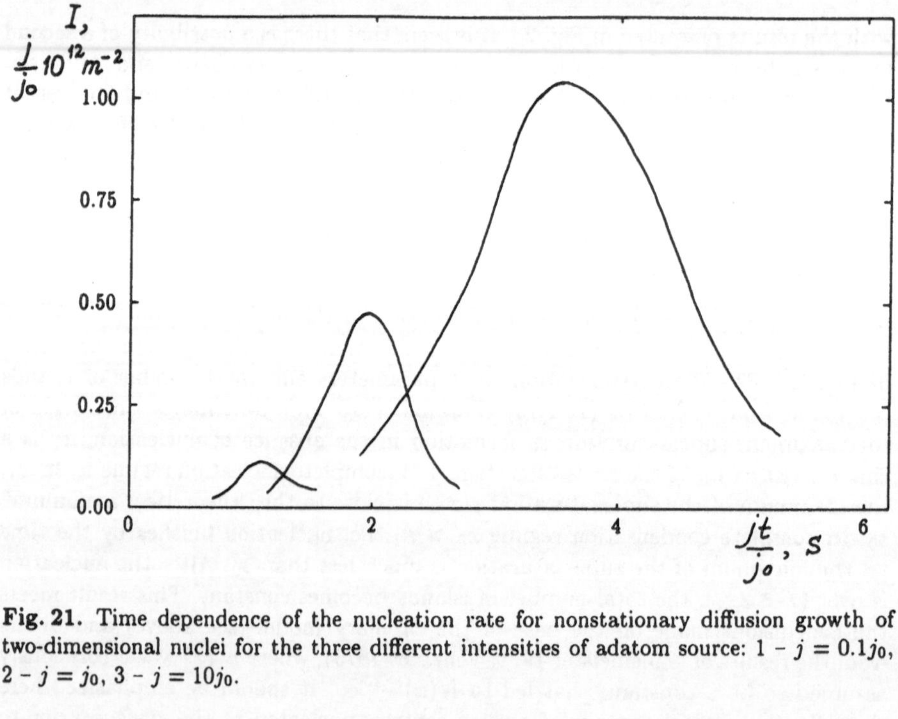

Fig. 21. Time dependence of the nucleation rate for nonstationary diffusion growth of two-dimensional nuclei for the three different intensities of adatom source: 1 - $j = 0.1 j_0$, 2 - $j = j_0$, 3 - $j = 10 j_0$.

during the whole process. The latter number tends obviously to the unity for $x \to \infty$, that corresponds to the nucleation of all the monomers in the system into one large cluster (monolayer).

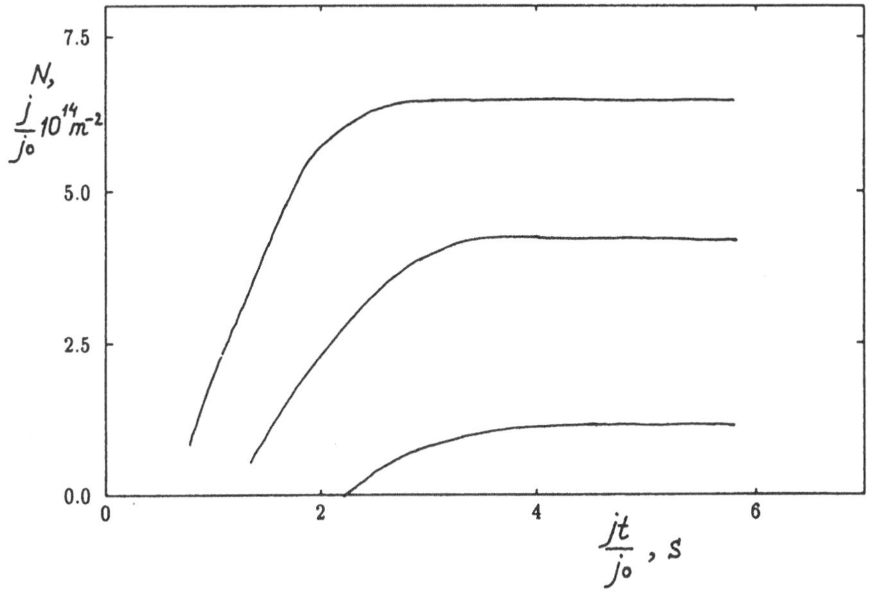

Fig. 22. Time dependence of the stable nuclei density for nonstationary diffusion growth of two-dimensional nuclei for the three different intensities of adatom source: $1 - j = 0.1j_0$, $2 - j = j_0$, $3 - j = 10j_0$.

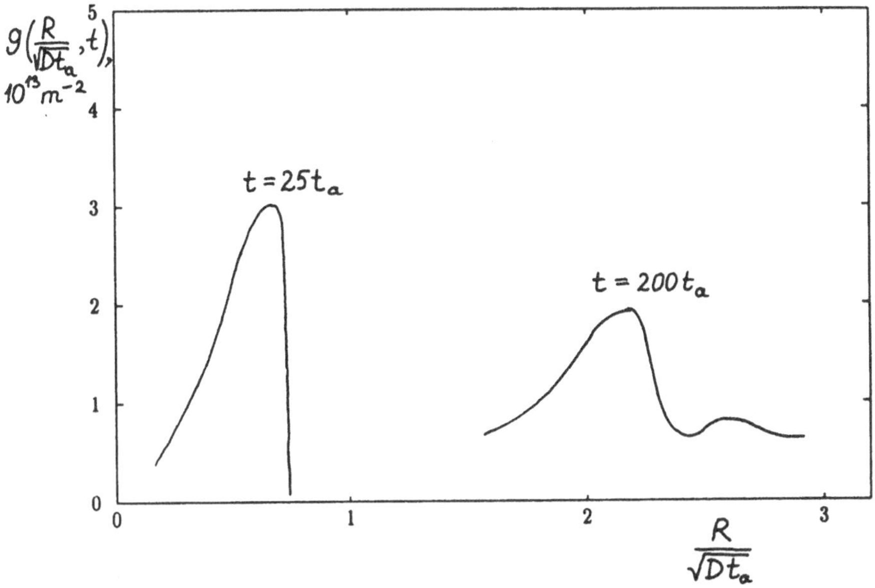

Fig. 23. Size distribution function of nuclei at the stage of coalescence.

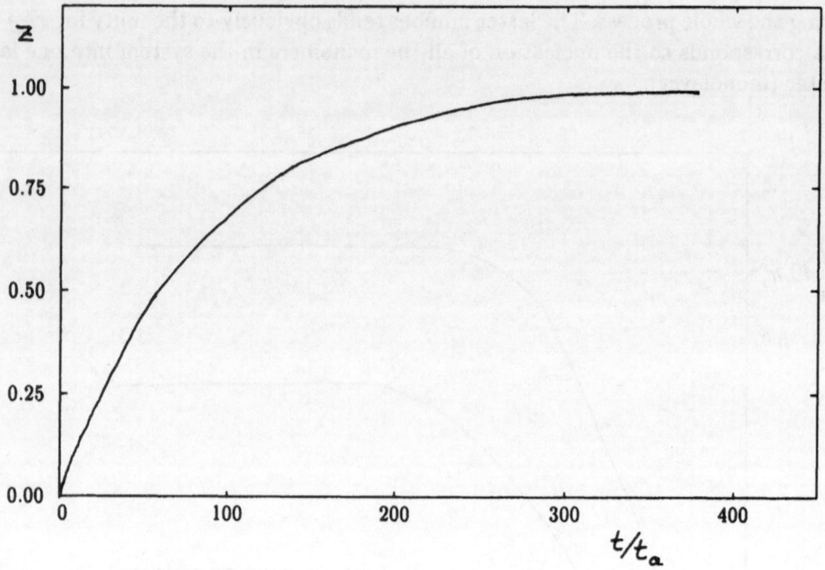

Fig. 24. Substrate coverage versus time dependence for exactly solvable model (Osipov 1990).

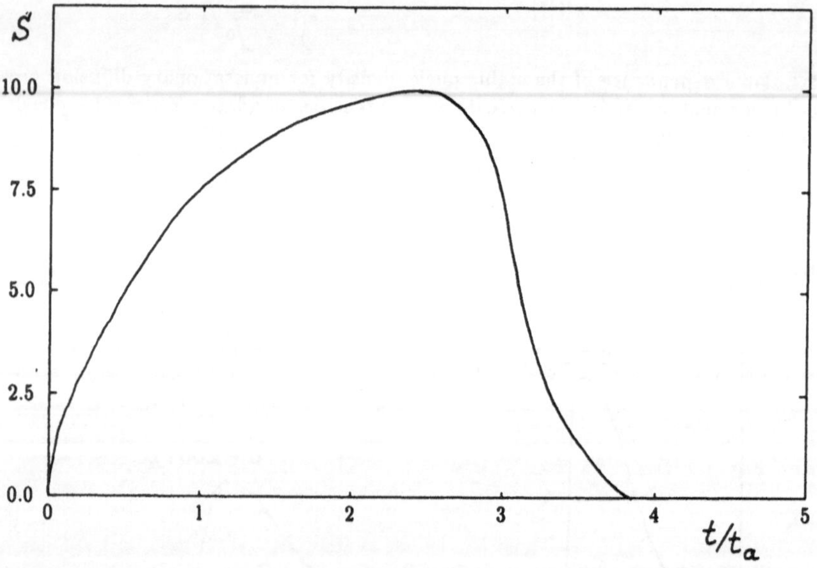

Fig. 25. Time dependence of the supersaturation in exactly solvable model (Osipov 1990).

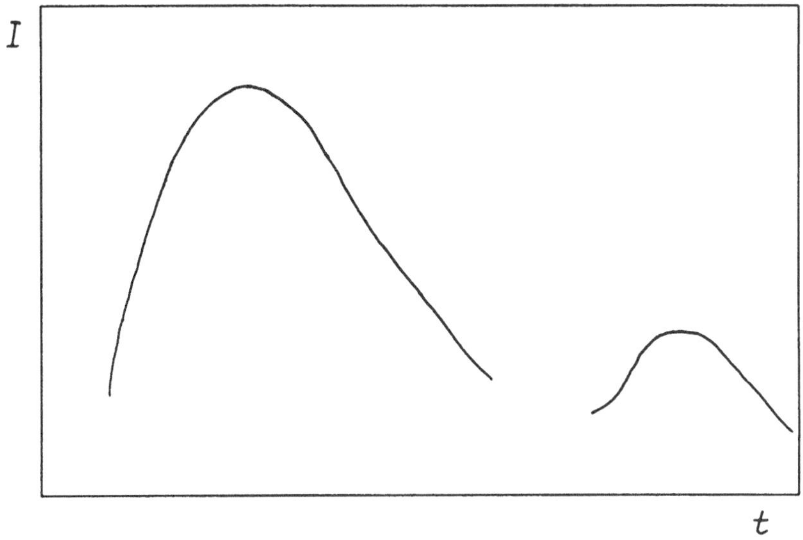

Fig. 26. Typical time dependence of I when v does not depend on i: the manifestation of the secondary nucleation phenomenon.

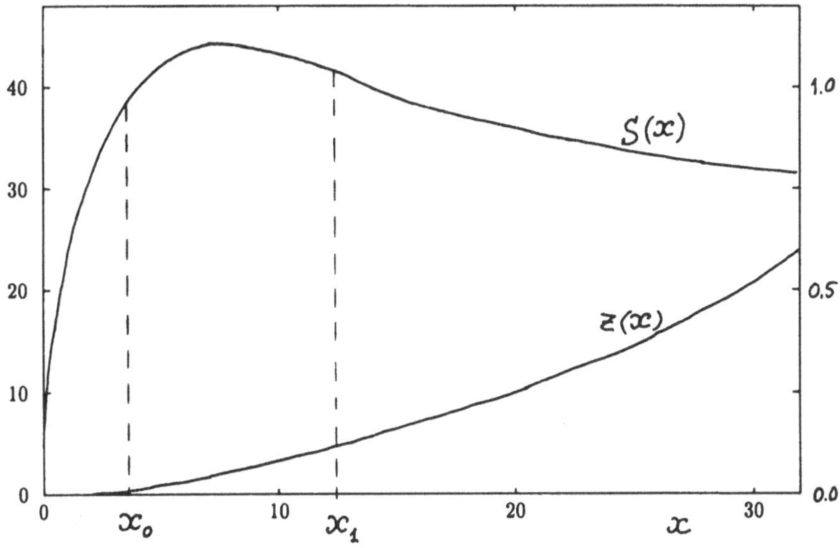

Fig. 27. Time dependences of the supersaturation s and the layer coverage z in the incomplete condensation regime ($A = 70$, $s_0 = 50$, $\kappa = 1$, $\theta = 0.0005$).

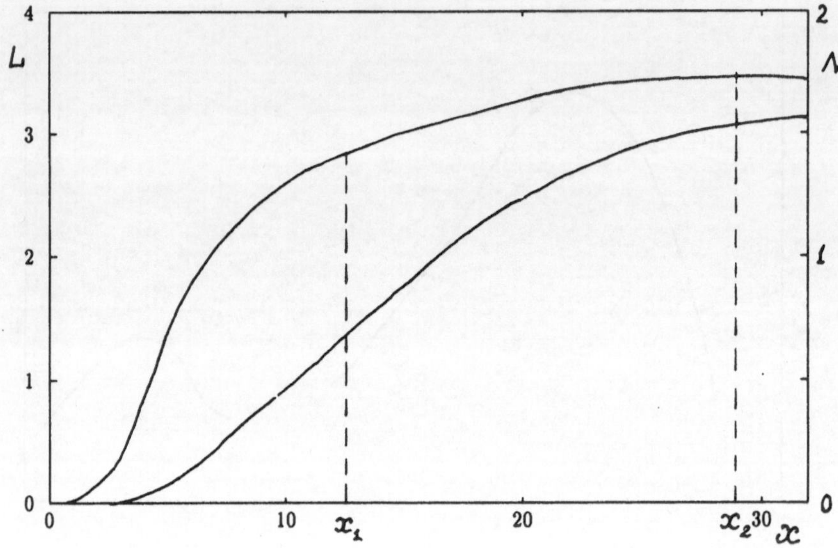

Fig. 28. Time dependence of the total number of clusters N and of the crystallization front perimeter L in the dimensionless units $2(D^R)^2 t_a^2 nN(x)$ and $D^R t_a L(x)/Cr$, respectively, for the same regime as in Fig. 27.

At the stage of direct coagulation the growth of perimeter of crystallization slows down and, thus, it is natural to define the characteristic time of continuos film formation begining as the maximum of $L(x)$ function: ($L'(x_2) = 0$) (Fig. 30,32).

The obtained numerical solutions show that the coagulation stage is usually separated from the nucleation one by the stage of the isolated growth ($x_1 \leq x \leq x_2$) when all the existing clusters grow through the absorption of adatoms and do not interact directly. Calculated curves of $s(x)$ meet qualitatively the experimental ones given in the review (Venables et al. 1984). Thus, the approach developed here allows one to introduce physical parameters to formal geometric model by Kolmogorov and to calculate both the size distribution function of nuclei $g(i)$ at the initial stage and integral morthological characteristics of the film. At the final stage of film formation, when description in terms of the distribution function $g(i)$ fails, the usage of variables $L(x)$, $z(x)$ remains adequate at arbitrary film coverages.

13.3 Multilayer Adsorption Film

On the base of the phenomenological model of multilayer film formation presented in Sect.12.2, the mean film height $\bar{h}(t)$ and corrugation have been calculated as functions of time from equations (12.2.19), (12.2.10). The obtained results have been compared to those of the computer simulation of A_3B_5 compounds growth in molecular beam epithaxy (Mayorov et al. 1988). The following model of growth has been used: the deposition from bicomponent source of As_2 and Ga onto the zink blende type substrate (001) face disoriented to (110) face by 0.95° has been

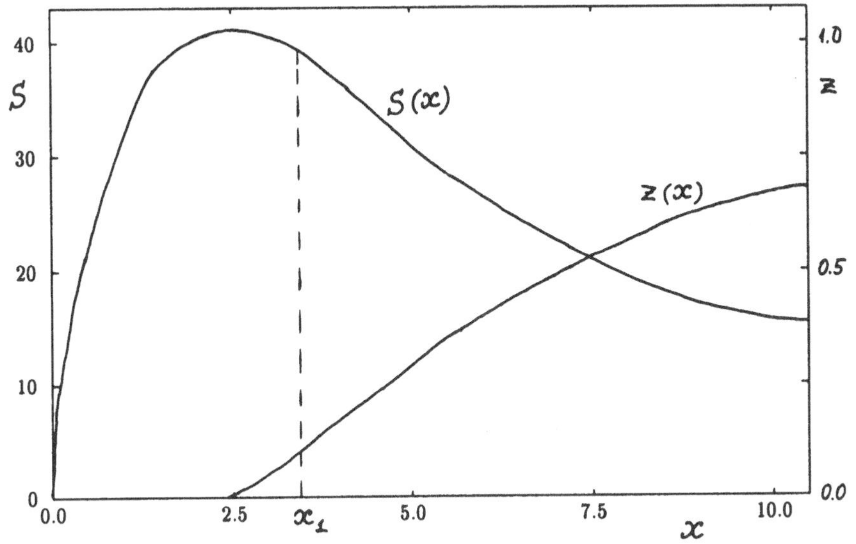

Fig. 29. Time dependences of the supersaturation s and the layer coverage z in the initially incomplete condensation regime ($A = 50$, $s_0 = 50$, $\kappa = 1$, $\theta = 0.001$).

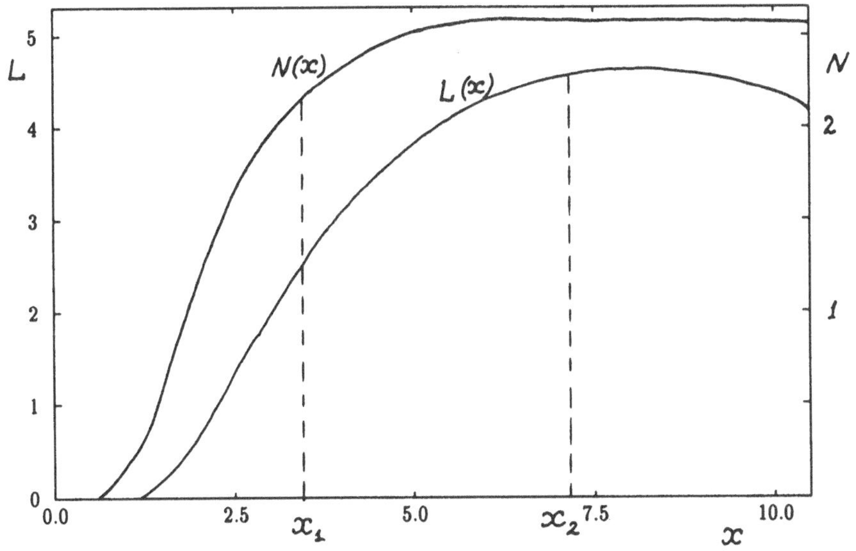

Fig. 30. Time dependence of the total number of clusters N and of the crystallization front perimeter L in the dimensionless units $2(D^R)^2 t_a^2 n N(x)$ and $D^R t_a L(x)/Cr$, respectively, for the same regime as in Fig. 29.

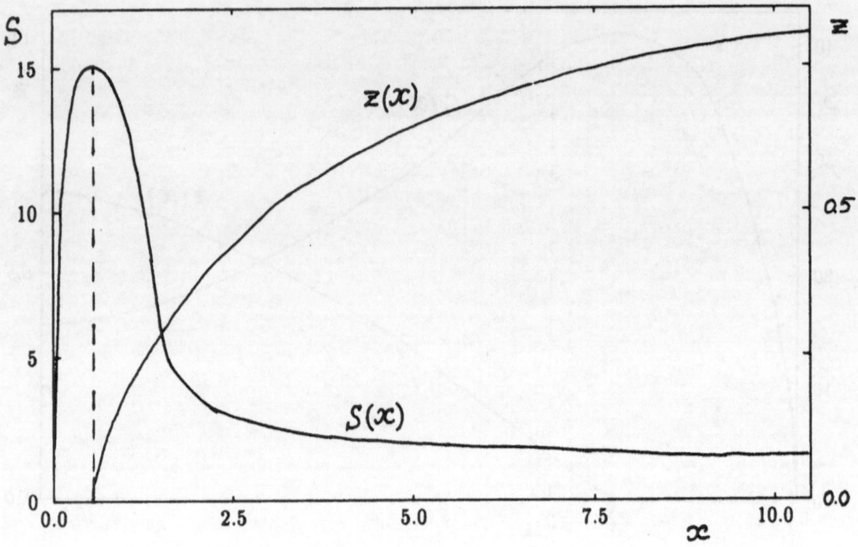

Fig. 31. Time dependences of the supersaturation s and the layer coverage z in the complete condensation regime ($A = 10$, $s_0 = 50$, $\kappa = 1$, $\theta = 0.001$).

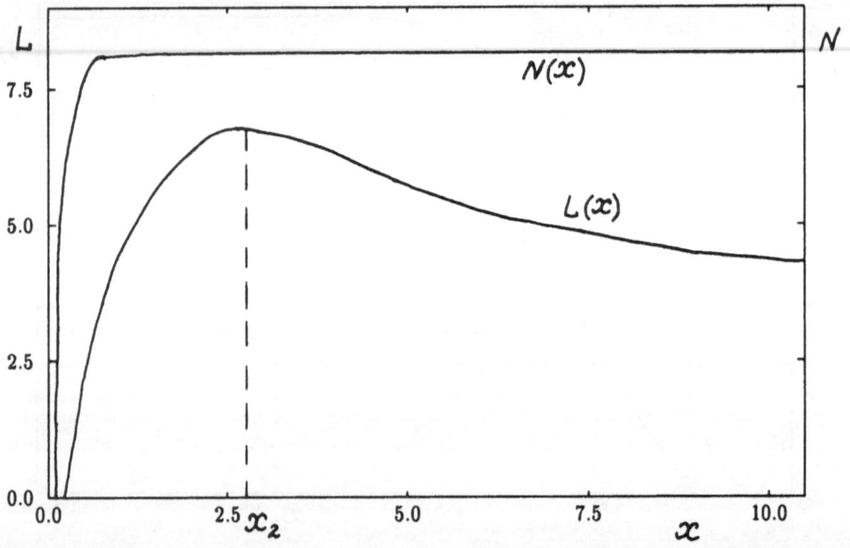

Fig. 32. Time dependence of the total number of clusters N and of the crystallization front perimeter L in the dimensionless units $2(D^R)^2 t_a^2 n N(x)$ and $D^R t_a L(x)/Cr$, respectively, for the same regime as in Fig. 31.

considered; the distance between steps (i.e. terrace width) has been assumed to be 120Å; surface reconstruction during the growth has been neglected; the dynamics of growth has been described with the help of model thermodynamical lattice gas Hamiltonian with an eye to the nearest and second neighbours; the probabilities of latteral diffusion and desorption have been taken in Gauss form with the activation barriers depending on the coverage of neighbouring sites. It is important to notice that the rate of growth of bicomponent A_3B_5 compounds has been defined only by metal flow (in the absence of Ga flow and with As source on, the growth did not occur). Therefore the dependence of the mean film height versus time was assumed to be analogous to that for unicomponent beams.

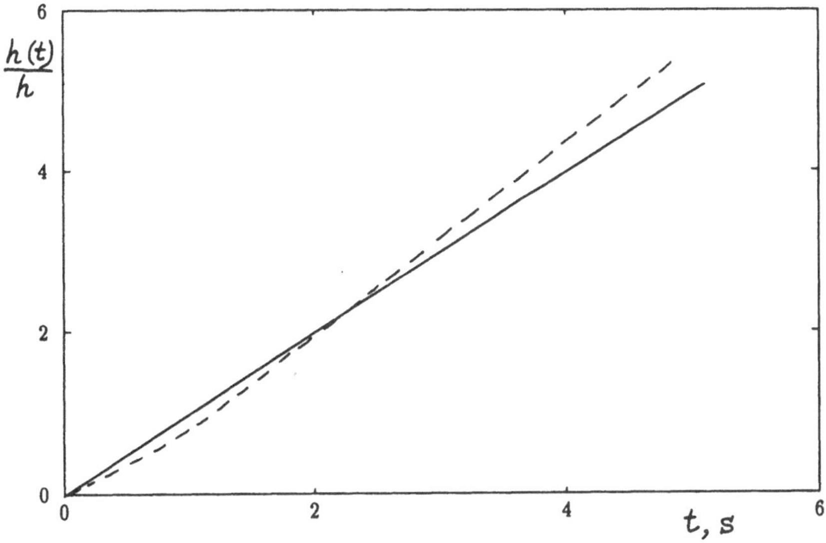

Fig. 33. Time dependence of the mean film height $\bar{h}$. Solid line presents the results of computer simulation of growth of GaAs compound (fluxes ratio $j_{As_2}/j_{Ga} = 3$, $T = 850\,K$). Dashed line – theoretical results obtained from (12.2.19) with $\nu = 1$, $F_2(x) = x^{m_2}$, $m_2 = 3$, $\omega_1 = 0.6\,s^{-1}$, $\omega_2 = 0.8\,s^{-1}$.

Figs. 33, 34 present the results of the comparison of thus simulated time dependencies of functions $\bar{h}(t)$ and $\Delta(t)$ with those obtained from (12.2.19), (12.2.20) using functions $A_{12}(x)$, $I_i(x)$, $K_i(x)$, and $B_i(x)$ in the form (12.2.16), (12.2.17). The $\bar{h}(t)$ curve has a shallow gap in the region of small t and then becomes linear. The plot of $\Delta(t)$ obtained via simulations has the form of square root function ($\Delta(t) \equiv t^{1/2}$) modulated with some oscillations disappearing for large times of deposition. The local minima of film corrugation (Fig. 34) correspond to the times of the first, the second and so on monolayers formation. The disappearance of the oscillations means that the layer-by-layer growth mechanism has been changed with the island one, and the formation of higher layers does not require the complete filling of underlying ones. The theoretical curve $\Delta(t)$ in Fig. 34 has only one local minimum. This is due to the assumption that the kinetics of higher layers formation (except for the first one) is similar, i.e. $F_1 \neq F_2 = F_3 = \ldots$, $\omega_1 \neq \omega_2 = \omega_3 = \ldots$. Using modified growth model with $F_1 \neq F_2 \neq \ldots \neq F_r = F_{r+1} = \ldots$, $\omega_1 \neq \omega_2 \neq \ldots \neq \omega_r = \omega_{r+1} = \ldots$ (r being the number of the last monolayer whose kinetics is affected by the substrate) leads to a curve with r minima (Kashchiev 1977). After the dissappearance of the oscillations ($t \gtrsim 3\,\mathrm{s}$, $\bar{h}(t) \geq 3h$, h - monolayer height) theoretical and simulated curves become identical. According to (12.2.23) the ratio of the squared corrugation to the mean height has asymptotic value $q_3(\infty) = 0.14$ ($m_2 = 3$). Therefore at $t = 4\,\mathrm{s}$ the film height is equal to 3.75 monolayers while the corrugation is of the order of 0.7 monolayers.

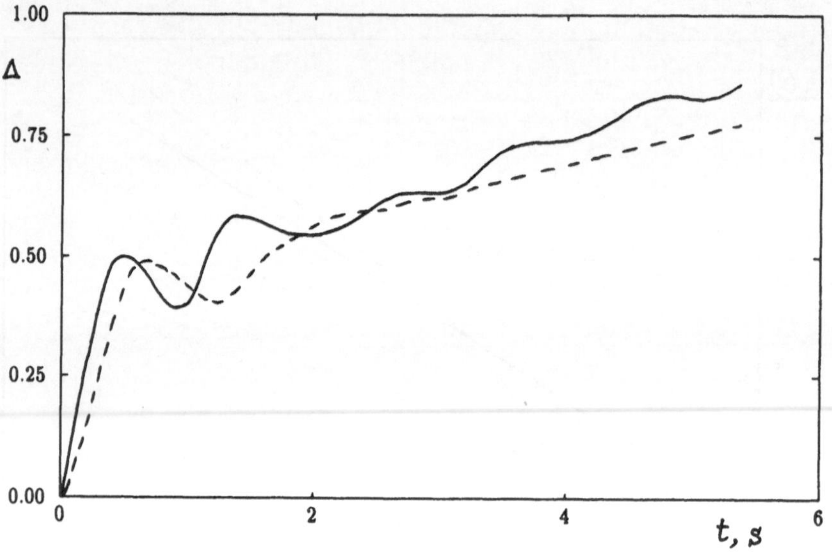

Fig. 34. Time dependence of the film height variance Δ. Solid line presents the results of computer simulation of growth of GaAs compound. Dashed line – theoretical results obtained from (12.2.19) under the same conditions as in Fig. 33.

Presented analysis allows one to make the following conclusions:

1. The simplest model of normal mechanism of layers formation from a unicomponent flow on the substrate leads to the Poisson distribution of relief points over the heights.
2. After filling a few first monolayers, the mean height and squared corrugation become linear functions of time. This is confirmed both by analytical investigations and computer simulations.
3. The film quality can be characterized by the parameter $q(\infty)$, that depends on the film growth mechanism. Therefore, the further development of multilayer films formation theory must be connected with the study of the detailed kinetics of processes in adsorption layers.

æ

IV

Kinetic Boundary Condition (KBC) and Inverse Scattering Problem

Introduction

The problem of kinetic boundary condition (KBC) was briefly discussed in Chap.6 in connection with the scattering problem. Here KBC will be considered in more detail. This problem is an important one providing a bridge between two different phases and taking into account effects of mutual interference of surface and gaseous systems.

Surface state (corrugation, thermal vibrations, catalyzing ability) influences the scattering of particles (electrons, neutrons, molecules) and determines the thermodesorption spectra features for real surfaces. The inclusion of these effects is necessary for the calculation of exchange and slip coefficients, for the study of Knudsen layer structure in the problems of aerodynamics and aerothermochemistry of flying vehicles, for surface and growing film diagnostics. On the other hand, the inclusion of gas phase influence on adsorbate is a question of great importance for the technological methods of thin film deposition from multicomponent multiphase sources (molecular beam epithaxy), for the chemical catalysis and material science. It is worth noting that conventional phenomenological KBCs do not take into account real surface processes and therefore can not be used for systematic description of interfaces. New approaches and results presented here allow one to formulate and solve the problems of the detailed kinetics of interphase interaction.

Nowadays there exist a number of different approximate solutions to boundary problem for molecular gas dynamics, that reduce it to integral brackets and exchange coefficients calculation. Therefore the problem of deriving adequate models of gas particles - real surfaces interaction and restoring their parameters becomes a most important one. It is this problem that constitutes the inverse gas–surface interaction problem.

The existence of three different levels of gas–surface interaction description allows one to point out three main types of the inverse problems:

1. Molecule - surface potential extraction from scattering information.
2. Potential parameters or scattering function extraction from exchange coefficients.
3. Local exchange coefficients or body geometry parameters extraction from total aerodynamical characteristics.

The first two types of the inverse problem will be considered in Chap.15 within a quasiclassical approximation.

14 KBC for Distribution Function in Gas Phase

14.1 General Representations of KBC

Let us discuss the general structure of the KBC (6.3.2) that includes processes of elastic and inelastic scattering, adsorption-desorption, and surface chemical reactions.

Using following representations

$$R_{c'c}(b',b) = [1 - a_{c'1}(b')]\tilde{P}_{c'c}(b',b), \quad D_{cf}(b) \equiv d_{cf}(b), \tag{14.1.1}$$

one can transform (6.3.2) to the form

$$p_z g_c^+(b)\Big|_{p_z>0} = \sum_{b'c'} |p'_z|(1 - a_{c'1}(b'))\tilde{P}_{c'c}(b',b) g_{c'}^-(b')\Big|_{p'_z<0} + d_{cf}(b). \tag{14.1.2}$$

The KBC presented by (14.1.2) is an inhomogenious one with the contributions from direct scattering $\tilde{P}_{c'c}(b',b)$ (including direct reactive scattering and scattering through bound states) and from desorption $d_{cf}(b)$ of particles (that have lost the correlation with the incoming flow due to relaxation in adsorbed state) being separated. Quantities $a_{c'1}(b')$ and $d_{cf}(b)$ are local adsorption and desorption coefficients, respectively. Indices 1 and f mean that an adsorption coefficient does not depend on $g_{c'}$ ($g_{c'} = 1$) and a desorption coefficient does depend on f_c. The inhomogenios KBC of (14.1.2) type is not used conventionally in practical calculations in the kinetic theory of gases. This condition can be transformed formally to a homogenious form

$$p_z g_c^+(b)\Big|_{p_z>0} = \sum_{b'c'} |p'_z| \tilde{K}_{c'c}(b',b) g_{c'}^-(b')\Big|_{p'_z<0}, \tag{14.1.3}$$

with the kernel $\tilde{K}_{c'c}(b',b)$ looking like

$$\tilde{K}_{c'c}(b',b|g_c f_c) = (1 - a_{c'1}(b'))\tilde{P}_{c'c}(b',b) + a_{c'1}(b')\tilde{d}_{cf}(b), \tag{14.1.4}$$

where

$$\tilde{d}_{cf}(b) = d_{cf}(b) \Big/ \sum_{b'c'} |p'_z| a_{c'1}(b') g_{c'}^-(b'). \tag{14.1.5}$$

As long as this kernel becomes a functional of g_c and f_c, the KBC becomes nonlinear with respect to the distribution function of gas particles g_c.

Local adsorption - desorption coefficients are related to the corresponding detailed probabilities via

$$a_{c'1}(b') = \sum_{Ec\beta} a_{c'c}(b', E\alpha), \quad d_{cf}(b) = \frac{1}{\sigma} \sum_{E'c',\beta'} \nu_{c'1} f_{c'}(E'\alpha') d_{c'c}(E'\alpha', b). \tag{14.1.6}$$

It is worth noticing here that local adsorption - desorption coefficients used in (6.1.11) had been defined through detailed probabilities $a_{c'c}$, $d_{c'c}$ in the following way

$$a_{cg}(E\alpha) = \sum_{b'c'} |v'_z| g_{c'}(b') a_c(b', E\alpha), \quad d_{c1}(E\alpha) = \sum_{b'} d_c(E\alpha, b'). \tag{14.1.7}$$

Separating the nonreactive and reactive mechanisms in direct scattering, adsorption and desorption events yields

$$\begin{aligned} \tilde{P}_{c'c}(b',b) &= \delta_{c'c}\tilde{P}_{c'}(b',b) + \tilde{P}^{\mathrm{r}}_{c'c}(b',b); \\ a_{c'c}(b',E\alpha) &= \delta_{c'c}a_{c'}(b',E\alpha) + a^{\mathrm{r}}_{c'c}(b',E\alpha); \\ d_{c'c}(E\alpha,b') &= \delta_{c'c}d_{c'}(E\alpha,b') + d^{\mathrm{r}}_{c'c}(E\alpha,b'); \end{aligned} \tag{14.1.8}$$

where index r denotes the probabilities of corresponding reactive transitions. In (6.1.11) all reactive transitions have been grouped in the operator X.

Let us define the fluxes of mass j_m, normal momentum j_z, tangential momentum j_τ, and energy j_E to the surface by the equations

$$j_l = j_l^- - j_l^+;\ \ j_l^\mp = \sum_{bc} |v_z| g_c^\mp(b) z_l, \quad v_z \lessgtr 0, \ l = m, z, \tau, E, \tag{14.1.9}$$

$$z_l = (m_c, p_z, p_\tau, \frac{p^2}{2m_c} + E_{\mathrm{int}}).$$

It is assumed here that $g_c^+(b)$ is obtained from KBC (14.1.3) and therefore all the flows are determined by the distribution function g_c^- near the surface and by the kernel $\tilde{K}_{c'c}(b',b)$. From the kinetic equation (6.1.11) one can obtain the system of transport equations for mass density (ρ_m), normal and tangential momenta (ρ_z, ρ_τ), and energy (ρ_E) in adsorbate. We shall use the following compact form of these equations

$$\partial_t \rho_l = \mathsf{G}_l(\rho_l) + j_l \tag{14.1.10}$$

with G_l being the operator describing transport of the l-th species due to relaxational processes at the substrate, j_l - is the flux of the l-th species to the surface from the gas phase. Equations (14.1.10) should be completed with transport equations in gas phase (e.g., Navier–Stokes equations). Thus, the system of equations (6.3.1), (6.1.11) together with KBC (14.1.3) allows one to investigate transport phenomena at interphase boundary on microscopic level, while equations (14.1.10) together with the transport equations in gas phase and momentum representation of KBC (14.1.3) allow one to do that on macroscopic level.

In the exact formulation these two approaches are closely connected with each other, while in approximate (thermodynamical or phenomenological) methods of interphase boundary description they may be considered independently for the generation of different models.

For example, the mass flux to the surface is described by

$$\begin{aligned} j_m &= \sum_{bc} |p_z| g_c^-(b)\Big|_{p_z<0} - \sum_{bc} d_{cf}(b) \\ &= \sum_{c,\beta\geq 1} (q(\theta)a_{cg*}(\alpha) - d_{c1*}(\alpha)\theta_c(\alpha)) \\ &= \sum_{c,\beta\geq 1} \mathsf{AD}_*(\theta_c, j_c). \end{aligned} \tag{14.1.11}$$

The condition of complete material balance between the gas phase and the adsorbate

$$j_m = 0 \tag{14.1.12}$$

leads to

$$\mathsf{AD}_*(\theta_c, j_c) = 0. \tag{14.1.13}$$

As previously shown, this equation in the absence of chemical reactions gives adsorption isotherm for immovable adsorbate ($\mathsf{G}_m(\rho_m) = 0$). For movable adsorbate ($\mathsf{G}_m(\rho_m) \neq 0$) the condition (14.1.13) gives partially equilibrium thermodynamic states, i.e., there is an equilibrium between the adsorbate and the gas phase, but the adsorbate itself is nonequilibrium. The reverse situation is possible, when $\mathsf{G}_m(\rho_m) = 0$ but $j_m \neq 0$. When complete material balance and chemical equilibrium over components at the boundary take place, one has to determine equilibrium values of θ_c, n_c from (14.1.13).

One can readily confirm that (14.1.12) leads to the following normalizing conditions on the probabilities in the kernel (14.1.4)

$$\sum_{bc} \tilde{P}_{c'c}(b'b) = 1, \quad \sum_{bc} \tilde{d}_{cf}(b) = 1, \quad \sum_{bc} \tilde{K}_{c'c}(b', b) = 1. \tag{14.1.14}$$

Vice versa, one can consider that the second condition in (14.1.14) is equivalent to (14.1.13) and is consistent with the adsorption isotherm, that can be obtained for chemically neutral gas from (14.1.13). It is obvious, that different assumptions on adsorbate nature lead to different conditions on the moments of the kinetic kernel $\tilde{K}_{c'c}$ and vice versa.

It is easy to demonstrate that making in (14.1.4), (14.1.5) a substitution $g_c \to g_{c0}$, $f_c \to f_{c0}$ (with $\tilde{d}_{cf} \to \tilde{d}_{cf0}$) and using microscopic detailed balance condition (of the type (8.2.16)–(8.2.19)) for $\tilde{P}_{c'c}$, $a_{c'c}$, $d_{c'c}$ yields, firstly, that the kernel (14.1.4) becomes independent on g_c, f_c (that means the linearization of KBC), and, secondly, that $\tilde{K}_{c'c}(b', b)$ satisfies microscopic detailed balance condition with the distribution function g_{c0} as well. It is known (Goodman and Wachman 1976), that the distribution function g_{c0} is the Maxwell-Boltzmann one and meets KBC (14.1.3). The solution to the Knudsen layer problem in this case may be found by means of conventional asymptotic methods (Cercignani 1975).

Considering a real surface, treated as a thermodynamical system, it is natural to postulate the existence of macroscopic detailed balance for averaged over energy probabilities (or for averaged adsorption and desorption coefficients) rather than the existence of microscopic detailed balance for the probabilities $a_{c'c}$, $d_{c'c}$. Under definite conditions, this leads to equations (14.1.12), (14.1.13), to correct isotherms, etc. In the framework of this approach one has to use moment methods (Cercignani 1975) for the solution of gas–surface interaction problem and to formulate both kinetic and macroscopic models. Such approach looks like a perspective one for the development of numerical and semianalytical methods of three - dimensional hypersonic flows calculations and for the calculation of highly anysotropic distributions of temperature and pressure fields near surfaces, in intermediate regimes between rarefied gas and continuous media, when the Knudsen number is of the order of unity. In this case model kinetic equations can be used as well as the kinetic boundary conditions, that are satisfied for example by anysotropic Maxwell - Boltzmann distribution function with different temperatures in tangential and normal to the surface directions. In such situation at some parts of flying object one has to use the description based on model kinetic equations and KBC, while at others – the description in terms of gasdynamic equations and corresponding hydrodynamic boundary conditions obtained from the kinetic theory.

14.2 Simplified Representations of KBC

Let us consider some simplified forms of KBC (14.1.3) with the kernel defined by (14.1.4).

As pointed out earlier, one can simplify expression (14.1.4) substituting functions g_c, f_c with equilibrium ones g_{c0}, f_{c0} along with the change $\tilde{d}_{cf} \to \tilde{d}_{cf0}$. This way of KBC linearization is valid provided the deviations of functions g_c, f_c from their equilibrium values are small. If it is not the case, one has to take into account the terms, arising from the expansion of the kernel $\tilde{K}_{c'c}$ over the small parameter. It is generally thought that for linearized KBC $\tilde{K}_{c'c}$ meets microscopic detailed balance principle.

In the framework of moment (or "model") method of the Knudsen layer problem solution it is natural to simplify the kernel (14.1.4) introducting the probabilities, assigning the weights to different channels of gas-surface interaction (nonreactive and reactive scattering, nonreactive and reactive adsorption, desoption, etc.), and detailed density distributions of particles in each channel over angles and energies. These channel weights can be found either from thermodynamical and phenomenological approaches or extracted from experiment, while detailed densities can be calculated from dynamical models of elementary processes. As an example, the following approximation of the kernel (14.1.4) can be used

$$\tilde{K}_{c'c}(b', b) = (1 - a_{c'0}) \left[s_{c'0}\delta_{c'c}\tilde{P}_{c'0}(b', b) + (1 - s_{c'0}) s_{c'c0}\tilde{P}_{c'c0}(b', b) \right]$$
$$+ a_{c'0}\left[d_{c'o}\delta_{c'c} + (1 - d_{c'0}) d_{c'c0} \right] \tilde{d}_{c0}(b), \qquad (14.2.1)$$

where $a_{c'0}$ and $d_{c'o}$ are integral adsorption and desorption coefficients, respectively, $s_{c'0}$ and $s_{c'co}$ are nonreactive and reactive direct scattering probabilities, $d_{c'co}$ is the reactive desorption probability. Index o indicates that equilibrium distribution functions are to be used when calculating a corresponding value. Symbols $\tilde{P}_{c'0}(b', b)$, $\tilde{P}_{c'c0}(b', b)$, and $\tilde{d}_{c0}(b)$ denote the detailed probabilities of direct nonreactive scattering, direct reactive scattering, and desorption, respectively. The condition of complete material balance (if valid) leads to the following relations

$$\sum_b \tilde{P}_{c'0}(b', b) = 1, \quad \sum_b \tilde{P}_{c'c0}(b'b) = 1, \quad \sum_b \tilde{d}_{c0}(b) = 1, \qquad (14.2.2)$$

$$\sum_c s_{c'c0} = 1, \quad \sum_c d_{c'c0} = 1. \qquad (14.2.3)$$

Detailed probabilities $\tilde{P}_{c'o}(b', b)$, $\tilde{P}_{c'co}(b', b)$, $\tilde{d}_{co}(b)$ are assumed to be calculated on the base of dynamical models, while average integral probabilities of adsorption, direct nonreactive and reactive desorption – on the base of thermodynamical and phenomenological approaches together with experimental data. For example, the adsorption probability may be presented as a sum of the probabilities related to reactive (r) and nonreactive (n) adsorption channels

$$a_{c'0} = a^{\rm r}_{c'0} + a^{\rm n}_{c'0}. \qquad (14.2.4)$$

The total reactive desorption probability may also be presented as a sum of the probabilities connected with different mechanisms (LH - Langmuir–Hinshelwood , D - direct):

$$d^{\rm r}_{c'c0} = d^{\rm D}_{c'c0} + d^{\rm LH}_{c'c0}, \qquad d^{\rm LH}_{c'co} = K_{c'co} d_{c0}, \qquad (14.2.5)$$

where $K_{c'c0}$ is the mean rate constant of a reactive transition $c' \to c$ at a surface.

For the case of nonreactive interaction one has to assume $s_{c'o} = d_{c'o} = 1$, that leads to the simplified form of (14.2.1)

$$\tilde{K}_{c'c}(b', b) = (1 - a_{co})\tilde{P}_c(b', b) + a_{co}\tilde{d}_{co}(b). \qquad (14.2.6)$$

This expression is a generalization of a widely used Maxwell scattering kernel that takes into account realistic inelastic scattering $\tilde{P}_c(b', b)$, desorption $\tilde{d}_{co}(b)$ and adsorption a_{co}.

14.3 Phenomenological Approximations for Boundary Operators

Here we shall review the most popular phenomenological approximations of KBC, that, though having no strict physical basis, have certain importance for applications. Such approximations contain fitting parameters and can be used for parametrisations of experimental results for realistic surfaces.

The most simple form of the scattering operator for structureless particles is given by the Maxwell kernel (Niven 1972) (it is common to use velocity variables $(\boldsymbol{v}, \boldsymbol{v}')$ instead of momentum ones $(\boldsymbol{p}, \boldsymbol{p}')$ for nonreactive scattering kernels)

$$\tilde{R}(\boldsymbol{v}', \boldsymbol{v}) = (1 - a_{10})\delta(\boldsymbol{v} - \boldsymbol{v}' + 2v_z{}'\boldsymbol{e}_z) + a_{10}\frac{v_z}{2\pi R_0^2 T_s}\exp\left(-\frac{v^2}{2R_0 T_s}\right), \qquad (14.3.1)$$

$$\boldsymbol{v}' = \boldsymbol{p}'/\mu, \quad \boldsymbol{v} = \boldsymbol{p}/\mu, \quad v_z{}' < 0, \quad v_z > 0, \quad R_0 = N_{\rm A} k_{\rm B}.$$

In the case of incomplete accomodation the temperеture of diffusively desorbing particles $T_{\rm d}$ would differ from that of the surface T_s. Energy accomodation coefficient α_E for such conditions is assumed to be of the form

$$\alpha_E = a_{10}(1 - \frac{4k_{\rm B}T_{\rm d}}{\mu v'^2})/(1 - \frac{4k_{\rm B}T_s}{\mu v'^2}). \qquad (14.3.2)$$

The Maxwell scattering kernel contains two dimensionless fitting parameters: a_{10} and $\gamma' = mv'^2/2k_{\rm B}T_s$, that can depend on velocity v' and surface features. Expression (14.3.1) meets both condition (14.2.1) and microscopic detailed balance. In (Epstein 1967) the following approximation for the local adsorption coefficient $a_1(\boldsymbol{v}')$ (that can be used instead of a_{10} in (14.3.1)) has been proposed

$$a_1(\boldsymbol{v}') = \exp\left(-b_1 v'^2\right) + a_{10}[1 - \exp\left(-b_2 v'^2\right)] \qquad (14.3.3)$$

with variable parameters b_1, b_2, and a_{10} to fit experimental data.

Another generalization of the model (14.3.1) is based on the introduction of the beam distribution indicatrix instead of specular part. The Maxwell or Gauss forms are used for this purpose. For example, in (Barantsev 1975) an expression of (14.3.1) type has been proposed for the description of the angular distribution of scattered particles

$$\tilde{R}(\Theta) = (1 - a_{10})(1 + \frac{2}{3}b_\omega)\frac{1}{\pi b_\omega^2} \exp\left(-\frac{(\Theta - \Theta_m)^2}{b_\omega^2}\right) \frac{\cos\Theta}{\cos\Theta_m} + a_{10}\cos\Theta, \quad (14.3.4)$$

where Θ is the reflection angle with Θ_m being its mean value. This model contains two independent parameters: a_{10} and Θ_m. All the rest parameters are expressed via these ones by virtue of the relations: $\eta = \Theta_m - \Theta'$, $b_\omega^2 = b_{\omega 0}^2 \cos^2\Theta'$, $\eta = \eta_0 \sin 2\Theta'$, $b_{\omega 0}^2 = 2\eta_0$.

Fixing impact velocity $\boldsymbol{v}' = \boldsymbol{V}$ yields the following relation between the reflection probability $\tilde{R}(\boldsymbol{v}', \boldsymbol{v})$ and the distribution function of reflected particles $g^+(\boldsymbol{v})$

$$\tilde{R}(\boldsymbol{v}', \boldsymbol{v}) = \frac{v_z}{\cos\Theta'} g^+(\boldsymbol{v}). \quad (14.3.5)$$

Nocilla proposed to approximate g^+ by the Maxwell function

$$g^+(\boldsymbol{v}) = n\left(\frac{\mu}{2\pi k_B T_s}\right)^{3/2} \exp\left(\frac{-\mu(\boldsymbol{v} - \boldsymbol{V})^2}{2k_B T_s}\right), \quad (14.3.6)$$

that leads to the following form of the indicatrix (the Nocilla model, see details in (Barantsev 1975)):

$$\tilde{R} = \frac{\cos\theta}{\pi} \exp\left(-\zeta^2 \sin^2\gamma\right) \frac{\chi(\zeta\cos\gamma)}{\chi(\zeta\cos\Theta)} \left[1 + \zeta\cos\gamma\Psi(\zeta\cos\gamma)\right], \quad (14.3.7)$$

with

$$\begin{aligned} \cos\gamma &= \cos\theta\cos\Theta \mp \sin\theta\sin\Theta\sin\Phi, \\ \chi(z) &= \exp(-z^2) + \sqrt{\pi}z(1 + \operatorname{erf} z), \\ \Psi(z) &= z + \frac{\sqrt{\pi}}{2\chi(z)}(1 + \operatorname{erf} z), \\ \zeta &= \sqrt{\frac{5}{6}}M, \end{aligned} \quad (14.3.8)$$

where θ and Φ are the spherical angles of vector $\boldsymbol{v}$, M is the Mach number. When $\zeta = 0$, (14.3.7) describes diffusive reflection, while for $\zeta \to \infty$ it gives a distribution concentrated in the direction $\theta = \Theta$ if $\Theta < 0$ and in the direction $\theta = \pi/2$ otherwise.

For the beam scattering description the so-called dispersion (or ray) model is used (Barantsev 1975), that is based on a combination of delta-functions. For example, for one–beam scattering it gives

$$\tilde{R}(\boldsymbol{v}', \boldsymbol{v}) = \delta(\boldsymbol{v} - \boldsymbol{v}_m(\boldsymbol{v}')), \quad (14.3.9)$$

with $\boldsymbol{v}_m(\boldsymbol{v}')$ being either taken equal to the mean velocity of peak-shaped distribution or extracted from macroscopic characteristics. The dispersion model uses the Gaussian functions with small dispersion instead of delta-functions. For example, in (Bartantsev 1975) the following approximation has been derived

$$\tilde{R}(v',v) = \frac{(1-3b_v^2/v_m^2)}{\sqrt{2\pi}b_v v_m^3}\tilde{R}_\omega v \exp\left(-\frac{(v-v_m)^2}{2b_v^2}\right), \tag{14.3.10}$$

$$\tilde{R}_\omega = \frac{v_m(\Theta,\Phi)\cos\Theta}{v_m(\Theta_m,\Phi_m)\cos\Theta_m}\left[1+2\frac{b_v^2}{v_m^2}-\chi(\Theta_m,\Phi_m)\right] \times \frac{1}{\pi b_\omega}\exp\left(-\frac{(\Theta-\Theta_m)^2}{b_\omega^2}\right), \tag{14.3.11}$$

$$\chi(\Theta,\Phi) = \frac{2b_v^2}{v_m^2}+\frac{b_\omega^2}{4v_m}\left[\frac{\partial^2 v_m}{\partial\Theta^2}+\frac{1}{\sin^2\Theta}\frac{\partial^2 v_m}{\partial\Phi^2}+(\cot\Theta-2\tan\Theta)\frac{\partial v_m}{\partial\Theta}-\frac{8}{3}v_m\right]. \tag{14.3.12}$$

With the help of the dispersion model one can readily calculate energy and momentum exchange coefficients and aerodynamical characteristics of different objects.

A mathematical model for the scattering indicatrix that meets the normalizing condition over final velocities and satisfies reciprocality principle has been proposed by Cercignani (Cercignani 1975). It contains accommodation coefficients of tangential momentum (α_τ) and normal energy (α_z) and has the form

$$\tilde{R}(v',v) = \frac{v_z}{\alpha_z\alpha_\tau(2-\alpha_\tau)2\pi(R_0T_s)^2} \times \exp\left(-\frac{v_z^2+(1-\alpha_z)v_z'^2}{2R_0T_s\alpha_z}-\frac{1}{\alpha_z(2-\alpha_\tau)}\frac{[v_\tau-(1-\alpha_\tau)v_\tau']^2}{2R_0T_s}\right) \times I_0\left(\frac{\sqrt{1-\alpha_z}v_zv_z'}{\alpha_zRT_s}\right) \tag{14.3.13}$$

with $I_0(x)$ being the modified Bessel function.

A simple analytical model of the scattering indicatrix has been derived in the framework of the eikonal approximation in (Bogdanov and Sergeev 1974). It has the Lorentzian form and contains two physical parameters ε and d

$$\tilde{R}(v',v) = \frac{2d/\pi E'}{\left[2(\sin\Theta'-\sin\Theta)+\frac{\varepsilon}{E'}\right]^2+d^2/E'^2}, \tag{14.3.14}$$

where Θ' and Θ are the incident and reflection angles, respectively, with ε and d representing the real and imaginary parts of the optical potential of gas-surface interaction. The parameter ε is a mean value of the interaction potential of particle and surface, while d includes provisional inelastic channels of scattering (energy exchange with phonon and electron subsystems, adsorbed layer, plasma vibrations, etc.). The mean value $\bar{V}$ of the potential is defined by the relation

$$\bar{V}\varepsilon r_0^2 \approx \left|\int_0^\infty dr r|V(r)|\exp\left(i\delta(r)\right)\right|, \tag{14.3.15}$$

$$\delta(r) = \int_0^\infty V(r(t))dt,$$

where $\delta(t)$ is the eikonal phase, r_0 is the potential radius. Potential $V(r)$ is called "soft" if the right hand side of (14.3.15) is defined for $E' \to \infty$, and it is called "rigid" otherwise. For energies $E' \approx V_0$ (V_0 is the potential well depth) using linear approximation of the potential one can obtain the relation $\varepsilon = (E' - V_0)/N$ with N being the number of particles in crystall block. For large E' one has to put $\varepsilon = E'/N$ for "rigid" potentials and $\varepsilon = V_0$ for "soft" ones. The parameter d varies significantly only in the vicinity of inelastis threshold, being rather flat beyond this region. Therefore it is useful to assume a stair-like approximation for d versus E' dependence, for example,

$$\begin{aligned} &d = V_0, \ E' < d_1; \\ &d = d_1, \ d_1 \le E' \le d_2; \\ &d = d_2, \ d_2 \le E' \le V_d; \\ &d = V_d, \ E' \ge V_d; \end{aligned} \tag{14.3.16}$$

with d_1 and d_2 being the thresholds of one and two phonons excitation, respectively, and V_d being surface sputtering threshold.

For small deviations of angles from the specular direction a simplified expression for the scattering indicatrix has been obtained by Sergeev (Sergeev 1978). This distribution is very close to Gaussian one. The comparison of the results of calculations for Ar–Ag(111) system on the base of this simplified formula with experimental results, presented in Fig. 35, seems satisfactory. The simplicity of the eikonal approximation makes it very convenient for applied aerodynamical investigations.

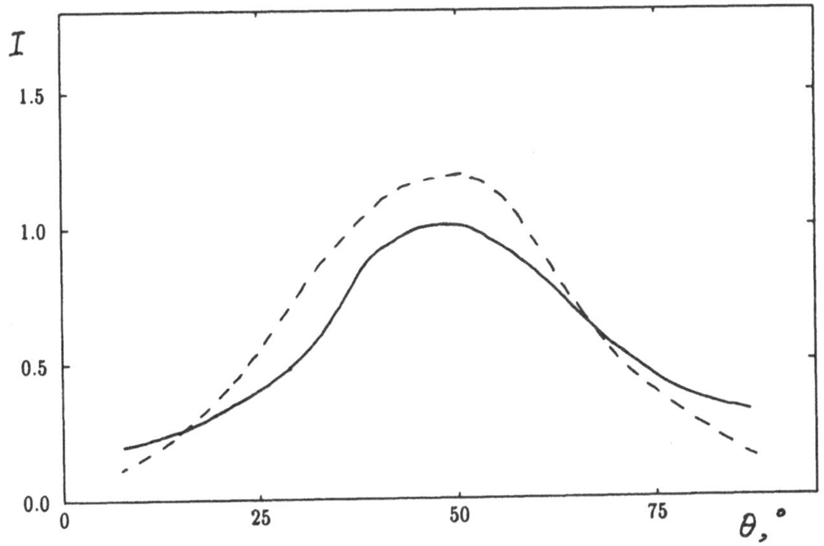

Fig. 35. The comparison of theoretical results (Sergeev 1978) (solid line) with experimental data (Hays et al. 1972) (dashed line): $\theta' = 58^o$, $E = 2,7$eV.

15 Inverse Scattering Problem in Gas–Surface Interaction

15.1 Gas–Cold Surface Interaction Potential Extraction: Quasiclassical Approach

As has been shown in Part I, the quasiclassical approximation leads to the separation of dynamical and statistical effects. This fact explains the significant advantage of this approach for the solution of inverse scattering problem.

The partial reflection probability to the Gth diffraction peak can be presented in the form $|T_{G,\Delta n}|^2 D(\Delta P, \Delta E; W)$, where $T_{G,\Delta n}$ is the diffraction scattering amplitude at a cold surface ($T_s \ll \Theta_s$) with the excitation of Δn quanta of a molecule, D is the dynamical structural factor of the surface vibrations, and W is the Debye–Waller exponent, that for fast collisions may be written as

$$W = \frac{1}{2}(\Delta p_z u)^2 \tag{15.1.1}$$

with u being the thermal vibration amplitude of surface atoms. A similar approximate representation with the Debye–Waller factor changed with the expression

$$W_a = \frac{1}{2}(\Delta P)^2 D_a \tag{15.1.2}$$

has been obtained for the case of a scattering from adsorbate layer (Bogdanov 1980). In the latter equation D_a is the diffusion coefficient of adatoms. Function D in this case is the Fourier-transform of the adatom–adatom correlation function.

In strict formulation the inverse gas–surface scattering problem is incorrect one. Therefore a number of further assumptions is to be done. Firstly, one has to postulate the potential form, e.g., a set of first order terms in the expansions (2.1.4), (2.1.12). Thus the potential extraction is reduced to the determination of the term amplitudes. Secondly, it will be postulated that a diffraction peak form does not depend on its number, so that a peak can be characterized entirely by its height

$$I_{G,\Delta n} \sim |T_{G,\Delta n}|^2. \tag{15.1.3}$$

This assumption is valid for low enough surface temperatures ($T_s \ll \Theta_s$), when the diffraction peak width is small. And, thirdly, the reflection probability is deemed to be a smooth function of level number.

Quasiclassical approximation for the amplitude $T_{G,\Delta n}$ results in the integral representation (3.2.5) involving the classical action increment. On the base of this representation and conventional expansion (Gel'fand and Shilov 1959)

$$\sum_{n=-\infty}^{+\infty} n^k \exp(inx) = 2\pi \sum_{n=-\infty}^{+\infty} \left[\frac{1}{i}\frac{\partial}{\partial x}\right]^k \delta(x - 2\pi n) \tag{15.1.4}$$

a method for the calculation of the amplitudes in the potential expansion from diffraction peaks heights has been developed (Bogdanov et al. 1986).

Let us introduce a compact notation for the following mean values

$$\langle\langle \mathcal{F}(G, \Delta n) | G, \Delta n \rangle\rangle \equiv \frac{\sum_{G,\Delta n} I_{G,\Delta n} \mathcal{F}(G, \Delta n)}{\sum_{G,\Delta n} I^0_{G,\Delta n}}, \tag{15.1.5}$$

$$\langle \mathcal{F}(G, \Delta n) | G \rangle \equiv \frac{\sum_G I_{G,\Delta n} \mathcal{F}(G, \Delta n)}{\sum_G I^0_{G,\Delta n}}. \tag{15.1.6}$$

Let us consider for the sake of simplicity a case of the scattering of a diatomic molecule (modelled by rotating oscillator) of mass m from a crystalline surface with lattice constants a_1, a_2. The first order terms of potential expansions (2.1.4), (2.1.12) have the form

$$V(\boldsymbol{r},\boldsymbol{q}) = V_0(z)\left[1+\kappa_1 \sin\frac{2\pi x}{a_1} + \kappa_2 \sin\frac{2\pi y}{a_2} + a_0^{l0} P_l(\cos\theta) + a_0^{01}(\rho-\rho_0)\right] \tag{15.1.7}$$

with $V_0(z)$ being the elastic potential, P_l - the Legendre polynomial ($l = 1$ for heteronuclear and $l = 2$ for homonuclear molecule); θ is the angle between the molecular axis and the surface normal, ρ - the interatomic distance in the molecule. To provide the possibility of consistent extraction of potential parameters we shall consider coefficients $\kappa_{1,2}$, a_0^{l0}, a_0^{01} to be constant. This is approved by the fact, that under quasiclassical conditions only small vicinity of turning point gives the main contribution to the action increment. With the potential (15.1.7) the action increment can be calculated in explicit form (Bogdanov et al. 1983)

$$\begin{aligned} \Delta S &= \Delta S_x + \Delta S_y + \Delta S_z + \Delta S_{\text{rot}} + \Delta S_{\text{vib}} \\ &= \left[\kappa_1 \hat{A}\sin\frac{2\pi x}{a_1} + \kappa_2 \hat{B}\sin\frac{2\pi y}{a_2} + \hat{C} + \hat{D}_0 + a_0^{l0}\hat{D}_l \sin l q_{\text{rot}} + a_0^{01}\hat{F}\sin q_{\text{vib}}\right], \end{aligned} \tag{15.1.8}$$

where q_{vib} and q_{rot} are the angle coordinates that correspond to the vibrational and rotational degrees of freedom, respectively, coefficients $\hat{A}$–$\hat{F}$ are expressed through the phase integrals over the classical trajectory and are determined by the elastic potential V_0. We do not need here the detailed expressions for these coefficients, but it is important that they can be calculated for definite models of elastic potential, do not depend on anisotropy coefficients, and weakly depend (through energy conservation law) on reciprocal lattice vector $\boldsymbol{G}$ and quantum numbers increment $\Delta\boldsymbol{n}$. Considering the multidimensional Fourier transform of the squared module of $T_{G,\Delta n}$ and using (15.1.7), (15.1.7), and (15.1.11), one can obtain by direct differentiation the following relations

$$\begin{aligned} \kappa_1^2\hat{A}^2 &= \frac{a_1^2}{2\pi^2}\langle\langle G_1^2|G,\Delta n\rangle\rangle = \frac{a_1^2}{2\pi^2}\langle G_1^2|G\rangle, \\ \kappa_2^2\hat{B}^2 &= \frac{a_2^2}{2\pi^2}\langle\langle G_2^2|G,\Delta n\rangle\rangle = \frac{a_2^2}{2\pi^2}\langle G_2^2|G\rangle, \\ (a_0^{l0})^2\hat{D}_l^2 &= \frac{1}{2\pi^2 l^2}\langle\langle \Delta n_{\text{rot}}^2|G,\Delta n\rangle\rangle = \frac{1}{2\pi^2 l^2}\langle \Delta n_{\text{rot}}^2|\Delta n\rangle, \\ (a_0^{01})^2\hat{F}^2 &= \frac{1}{2\pi^2}\langle\langle \Delta n_1^2|G,\Delta n\rangle\rangle = \frac{1}{2\pi^2 l^2}\langle \Delta n_1^2|\Delta n\rangle. \end{aligned} \tag{15.1.9}$$

These formulae allow one to determine the anisotropy parameters from experimentally measured diffraction peak heights $I_{G,\Delta n}$. The anisotropy parameters of higher orders in general expansion of the potential can be extracted in the same manner.

It is worth emphasizing that formulae (15.1.9) allow one to determine potential parameters from both complete and incomplete experimental data. This is due to the fact, that under used approximation to the dynamical problem solution, contributions of particular inelastic processes are factorized. The using of realistic classical trajectories will abandon this property: only the first equations in (15.1.9) will be valid with the redefined coefficients $\hat{A}$–$\hat{F}$.

Let us illustrate the method of potential parameters $\kappa_{1,2}$ extraction from experimental results on the example of the system He–LiF(001). In (Goodman and Wachman 1976) the experimental heights of 64 diffraction peaks have been presented with the surface temperature being $T_s = 10\,\mathrm{K}$ and wave vector of normally incident beam - $k = 10.38\,\mathrm{\AA}^{-1}$. The Morse potential will be used for modelling elastic interaction

$$V_0(z) = D\left(e^{-2\lambda z} - 2e^{-\lambda z}\right) \tag{15.1.10}$$

with the following parameters (Goodman and Wachman 1976) $D = 176\,\mathrm{cal/mol}$, $\lambda = 1.1\,\mathrm{\AA}^{-1}$. We shall take into account five peaks lying in the incidence plane ($\phi = 0$), beginning with the specular one. The values of the coefficient $\hat{A}$ for different peaks are (Bogdanov et al. 1986): $\hat{A}(0,0) = 4.94$, $\hat{A}(1,0) = 5.59$, $\hat{A}(2,0) = 5.71$, $\hat{A}(3,0) = 4.75$, $\hat{A}(4,0) = 3.40$. This sequence of values confirms that $\hat{A}$ versus diffraction peak number dependence is really weak enough and above done assumption is valid. Using in (15.1.9) the mean value $\hat{A} = 4.88$ and adopting for the definity $I_{(0,0)} = I_{(1,0)}$ (in (Goodman and Wachman 1976) the specular peak scale differs from that of nonspecular ones), yields anisotropy parameter $\kappa_1 = 0.46$. Thus, taking into account that for LiF(001) $a_1 = a_2 = a = 2.84\,\mathrm{\AA}$, gives one the following interaction potential

$$V(x,y,z) \approx V_0(z)\left[1 + 0.46\sin\frac{2\pi x}{a} + 0.46\sin\frac{2\pi y}{a}\right] \tag{15.1.11}$$

with $V_0(z)$ being defined by (15.1.10).

Thus obtained anisotropy parameter κ_1 may be compared with that from (Goodman and Wachman 1976), where it was obtained using slightly different model of potential. Namely, $V_0(z)$ was taken to have the Morse form, while diffractive potentials $V_G(z)$, $|G| > 0$ were assumed there to be exponentially repulsive ones

$$V_G(z) = \kappa_G D e^{-2\lambda z}, \tag{15.1.12}$$

where the value of κ_{G_0} for the minimal nonzero vector of the reciprocal lattice $G_{\min} = G_0$, $G_0 = 2\pi/a = 2.21\,\mathrm{\AA}^{-1}$ was equal to 0.10–0.11. For such potential the following expression is an analog to the anisotropy paramrter of our model:

$$\frac{2\kappa_{G_0}}{1 - 2e^{\lambda z_*}}$$

with z_* being the turning point for the energy of He beam $E = 750\,\mathrm{K}$. Simple calculations show that this value of κ_{G_0} corresponds to $\kappa_1 = 0.40$–0.44 that is in a good agreement with our value of $\kappa_1 = 0.46$.

As a second example let us extract parameter a_0^{l0} for the system CO–Pt, where rainbow effects in vibrational-rotational transitions have been observed. In experiments (Mantell et al. 1983) on the scattering of CO ($E \approx 300\,\mathrm{K}$) from thermal surface of Pt ($T_s = 1365\,\mathrm{K}$) they have observed two vibrational transitions $n_{\mathrm{vib\,i}} = 1 \to n_{\mathrm{vib\,f}} = 0$ and $n_{\mathrm{vib\,i}} = 2 \to n_{\mathrm{vib\,f}} = 1$ with multiple rotational sattelites. Taking the Born-Mayer potential as a model for V_0

$$V_0(z) = A\exp(-\lambda z) \tag{15.1.13}$$

with $\lambda = 1.91\,\text{Å}^{-1}$ and performing analogous calculations (see (Bogdanov et al. 1986) for details) show that parameter $\hat{D}_1$ in the action increment (15.1.8) weakly depends on quantum numbers $n_{\text{rot i}}$, $n_{\text{rot f}}$ so that one can use a mean value $\hat{D}_1 \approx 20$ for the registered transitions n_{rotf} =1–20. Finally, using this value in (15.1.9) yields for the rotational anisotropy parameter $a_0^{l0} = 0.08$. Experimental data (Mantell et al. 1983) are insufficient for the extraction of the vibrational anisotropy parameter a_0^{01}, because of $\Delta n_{\text{vib}} = n_{\text{vib i}} - n_{\text{vib f}} = 1$ for the both registered transitions.

15.2 Extraction of the Interaction Potential from Exchange Coefficients: Quasiclassical Approach

The molecular beam experiments on rainbow scattering are rather expensive. Therefore the efficiency of the gas–surface interaction is described frequently in terms of the exchange coefficients of momentum

$$\pi = \frac{p_{\text{i}} - \langle R p_{\text{f}} \rangle}{p_{\text{i}}} \cos \Theta \tag{15.2.1}$$

and energy

$$\epsilon = \frac{p^2 - \langle R(p_{\text{f}}^2) \rangle}{p_{\text{i}}^2} \cos \Theta , \tag{15.2.2}$$

where

$$\langle R f(p_{\text{f}}, n_{\text{f}}) \rangle \equiv \sum_{n_{\text{f}}} \int_{p_{\text{f}z}>0} \mathrm{d}p_{\text{f}} R f(p_{\text{f}}, n_{\text{f}}), \tag{15.2.3}$$

with R being the scattering indicatrix (3.1.1), normalized by unity.

Considering for the sake of simplicity the case of structureless particle scattering from thermal vibrations of surface and reducing the influence of inelastic processes to the thermal attenuation of diffractive peaks, one can present the scattering kernel in the form

$$R(p_{\text{f}}, p_{\text{i}}) = \frac{p_{\text{f}z}}{\mu} \sum_{G} \delta(\Delta P - G) \delta(\Delta E) \mathrm{e}^{-2W(G)} J_n^2(\kappa_1 \hat{A}) J_m^2(\kappa_2 \hat{B}). \tag{15.2.4}$$

This reduction is valid in the case of a cold surface ($T_s < \Theta_s$). Approximating the Bessel functions in the scattering kernel by the first non-vanishing terms of their Taylor expansion gives one an asymptotic expansion of I_G over adiabatic parameter (slipping incidence angles $\Theta \to \pi/2$ need special consideration and are excluded here) for an arbitrary form of the coefficients $\hat{A}$ and $\hat{B}$, i.e. for an arbitrary form of the elastic potential $V_0(z)$ in (15.1.7). We shall use the Born–Mayer potential (15.1.19) to analyze this expansion getting in the lowest non-vanishing approximation (Tiganov 1987)

$$\pi_1 = \frac{2\kappa_1^2}{\pi^2} \frac{\cos^2 \Theta}{\Omega_1 S_1^2 T_1} \left[\Omega_1^2 + \sigma_1^2 (1 - T_1) - 2 W_0 \sigma_1 \Omega_1^2 T_1^2 \right] + \mathcal{O}(\kappa_1^4), \tag{15.2.5}$$

$$\pi_2 = -\frac{2\kappa_2^2}{\pi^2} \frac{\cos^2 \Theta}{\Omega_2 S_2^2 T_2} \left[\Omega_2^2 + \sigma_2^2 (1 - T_2) - 2 W_0 \sigma_2 \Omega_2^2 T_2^2 \right] + \mathcal{O}(\kappa_2^4), \tag{15.2.6}$$

$$\pi_z = 2\cos^2 \Theta \left\{ 1 + \frac{\kappa_1^2}{\pi^2 S_1^2 T_1} \left[(\Omega_1^2 + \sigma_1^2)(\frac{1}{\sigma_1} - T_1) - \sigma_1 T_1 - 2 W_0 \Omega_1^2 T_1 \right] \right.$$

$$+\frac{\kappa_2^2}{\pi^2 S_2^2 T_2}\left[(\Omega_2^2+\sigma_2^2)(\frac{1}{\sigma_2}-T_2)-\sigma_2 T_2-2W_0\Omega_2^2 T_2\right]\Big\}+\mathcal{O}((\kappa_1^2+\kappa_2^2)^2). \tag{15.2.7}$$

In (15.2.5)–(15.2.7) the following notation is used ($i = 1, 2$):

$$S_i=\frac{\sinh\Omega_i}{\Omega_i},\qquad T_i=\frac{\tanh\Omega_i}{\Omega_i},\qquad \Omega_i=\frac{\sigma_i p_i}{|p_i|},\qquad \sigma_i=\frac{2\pi^2}{a_i\lambda}, \tag{15.2.8}$$

and W_0 is the Debye–Waller exponent for the specular peak ($G = 0$).

Provided the value of λ in (15.1.13) is known one can easily extract the anisotropy parameters $\kappa_{1,2}$ from $\pi_{1,2}$ by means of equations (15.2.5), (15.2.6). In more general case, equations (15.2.5)–(15.2.7) may be considered as a system of equations on three potential parameters: λ, κ_1 and κ_2, where W_0 should be defined from independent experiment.

Thus, in the framework of the quasiclassical approach the ansatz on interaction potential form, together with the structure of a surface under consideration, allows one to extract by simple means the potential parameters from experimental data on different levels of description. It is worth noting, that despite quasiclassical theory in some cases allows one to extract only characteristic combinations of physical parameters, it is these combinations that are contained in the inelastic scattering probabilities and, therefore, govern the efficiency of momentum and energy exchange in the gas–surface scattering. æ

References

Akhiezer, A.I., Peletminskiy, S.V. (1977): "Methods of Statistical Physics" (Nauka, Moscow) (in Russian)

Armand, G., Salanon, B. (1987): Surf. Sci. **217** 341

Asada, H. (1990): Surf. Sci. **227** L125

Balakhonov, N.F., Zak, D.I. (1988): "Boundary Conditions for the Boltzmann Equation and Thermalization of Particle Flows at Surface", in Proc. 9th National Conf. Rarefied Gas Dyn. (Moscow Power Inst., Moscow) Vol. 2, pp. 44–48 (in Russian)

Barantsev, R.G. (1975): "Interaction of Rarefied Gases with Surfaces" (Nauka, Moscow) (in Russian)

Bauer, E., Green, A.K., Kuntz, K.M., Poppa, H. (1966): "The Formation of Thin Continuous Films From Isolated Nuclei", in Proc. of Int. Conf. on Basic Problems in Thin Film Physics (Göttingen) pp. 135–152

Beeby, J.L. (1971): J. Phys C **4** L359

Belen'kiy, V.Z. (1980): "Geometrical - Probabilistic Models of Crystallization" (Nauka, Moscow) (in Russian)

Berry, M.V. (1975): J. Phys. A **8** 566

Billing, G.D. (1975): J. Chem. Phys. **62** 1480

Billing, G.D. (1990): Comput. Phys. Rep. **12** 383

Binder, K., Ed.(1979): "Monte Carlo Methods in Statistical Physics" (Springer Verlag, Berlin e. a.)

Blackman, M., Guzzon, A.E. (1959): "On the Size Dependence of the Melting and Solidification Temperatures of Small Particles of Tin", in: "Structure and Properties of Thin Films" (John Wiley, New York) pp. 217–220

Blinov, N.V., Kulginov D.V. (1991): Poverkhnost (USSR), No. 11 5 (in Russian)

Bogdanov, A.V. (1980): "Semiclassical Representations in the Problem of Gas–Surface Interaction", in Goodman, F.G., Wachman, H.Y.: "Dynamics of Gas–Surface Scattering", Russian ed. (Mir, Moscow), pp. 373–405 (in Russian)

Bogdanov, A.V., Sergeev, V.L. (1974): "Indicatrix of Gas Scattering at Crystalline Surface", in: "Aerodynamics of Rarefied Gases" (Leningrad State Univ., Leningrad), No.7, p. 71 (in Russian)

Bogdanov, A.V., Gorbachev, Yu.E., Strelchenya, V.M. (1983): "Rotational Excitation of Diatomics and Polyatomics in Gas–Surface Scattering", Preprint No. 839 of Ioffe Physical–Technical Institute (Leningrad)

Bogdanov, A.V., Dubrovskiy, G.V., Fedotov, V.A. (1985a): Zh.Fiz.Khim. (USSR) **49** 1219 (in Russian)

Bogdanov, A.V., Gorbachev, Yu.E., Strelchenya, V.M. (1985b): Poverkhnost (USSR), No. 1 45 (in Russian)

Bogdanov, A.V., Nguen Nam An, Tiganov, I.I. (1986): Poverkhnost (USSR), No. 11 35 (in Russian)

Bogdanov, A.V., Dubrovskiy, G.V., Gorbachev, Yu.E., Strelchenya, V.M. (1989): Phys. Rep. **181** 121

Bogdanov, A.V., Dubrovskiy, G.V., Osipov, A.I. and Strelchenya, V.M. (1991): "Rotational Relaxation in Gases and Plasma" (Energoatomizdat, Moscow) (in Russian)

Borisov, S.F., Balakhonov, N.F., Gubanov, V.A. (1988): "Interaction of Gases with Solid Surfases" (Nauka, Moscow) (in Russian)

Borman, V.D., Krylov, S.Yu., Prosyanov, A.V. (1988): Zh. Eksp. Teor. Fiz. (USSR) **94** 271 (in Russian)

Bortolani, V., Franchini, A., Nizzoli, F., Santoro, G. (1983): Surf. Sci. **128** 249

Brako, R., Newns, D.M. (1982): Surf. Sci. **117** 42

Brivio, G.P., Grimley, T.B. (1979): Surf. Sci. **89** 226

Celli, V., Himes, D., Bortolani, V., Santoro, G., Toennis, J.P., Zhang, G. (1991): Surf. Sci. **242** 518

Cercignani, C., (1975): "Theory and Application of Boltzmann Equation" (Scotish Academic Press, Edinburgh – London)

Ceyer, S.T. (1988): Ann. Rev. Phys. Chem. **39** 479

Chung, S., Holter, N., Cole, M.W. (1985): Phys. Rev. B **31** 6660

Chung, S., Holter, N., Cole, M.W. (1986): Surf. Sci. **165** 466

Croxton, C.A. (1974): "Liquid State Physics – A Statistical Mechanical Introduction" (Cambridge Univ. Press, Cambridge)

Dash, J.G. (1975): "Films on Solid Surfaces" (Academic Press, New York)

Doll, J.D., Voter, A.F. (1985): J.Chem.Phys. **82** 80

Dubrovskiy, G.V. (1982): Zh. Tekhn. Fiz. (USSR) **52** 1927 (in Russian)

Dubrovskiy, G.V. (1991a): "Adsorption, Catalytical Reactions and Their Influence on Aerodynamical Characteristics", in Proc. 10th National Conf. Gas Dyn. (Moscow Power Inst., Moscow) Vol. 2, pp. 10–17 (in Russian)

Dubrovskiy, G.V. (1991b): "Microscopic Models of Heat and Mass Transfer During Adsorption Film Growth", in Proc. of Int. School-Seminar on Kinetic Theory of Transport Processes in Evaporation and Condensation (Lykov Heat and Mass Transfer Inst., Minsk) pp. 38–50 (in Russian)

Dubrovskiy, G.V., Bogdanov, A.V. (1979a): Chem. Phys. Lett. **62** 89

Dubrovskiy, G.V., Bogdanov, A.V. (1979b): Zh. Tekhn. Fiz. (USSR) **49** 1386 (in Russian)

Dubrovskiy, G.V., Zyryanov, V.V. (1987): Vestnik LGU (USSR) Ser. I No. 15 105 (in Russian)

Dubrovskiy, G.V., Bogdanov, A.V., Gorbachev, Yu.E., Strelchenya, V.M. (1983): Poverkhnost (USSR), No. 3 56 (in Russian)

Dubrovskiy, G.V., Zyryanov, V.V., Fedotov, V.A. (1988): "Physical Models of Adsorption and Catalytical Reactions at Surface", in Proc. 9th National Conf. Rarefied Gas Dyn. (Ural State Univ., Sverdlovsk) Vol. 2, pp. 3–7 (in Russian)

Dubrovskiy, V.G. (1990): "Kinetic Models of Cluster and Thin Film Growth": Ph. D. Thesis (St.Petersburg) (in Russian)

Epstein, M. (1967): AIAA J. **5** 1797

Faddeev, L.D., Slavnov, A.A. (1980): "Gauge Fields. Introduction to Quantum Theory" (Benjamin, New York)

Fedotov, V.A. (1988): "To the Theory of Phonon Scattering of Diatomic Gas", in Proc. 9th National Conf. Rarefied Gas Dyn. (Ural State Univ., Sverdlovsk) Vol. 2, pp. 37–44 (in Russian)

Ferziger, J.H., Kaper, H.G. (1972): "Mathematical Theory of Transport Processes in Gases" (North-Holland, Amsterdam)

Feynman, R.P., Hibbs, A.R. (1965): "Quantum Mechanics and Path Integrals" (McGraw-Hill, New York)

Flood, E.A., Ed. (1967): "The Solid – Gas Interface", Vol. 1 (Marcel Dekker Inc., New York)

Gel'fand, J.M., Shilov, G.E. (1959): "Generalized Functions" (Fizmatgiz, Moscow) (in Russian)

Gel'fand, J.M., Yaglom, A.M. (1960): J. Math. Phys. **1** 48

Gibson, K.D., Sibener, S.J. (1988): J. Chem. Phys. **88** 7862

Glauber, R.I. (1963): J. Math. Phys. **4** 294

Goodman, F.O, Wachman, H.Y. (1976): "Dynamics of Gas–Surface Scattering" (Academic Press, New York)

Gorbachev Yu.E., Strelchenya V.M., Fedotov V.A. (1991): Poverkhnost (USSR), No. 2, 5 (in Russian)

Grabert, H., Schramm, P., Ingold, G.-L. (1988): Phys. Rep. **168** 115
Hays, W.J., Rodgers, W.E., Knuth, E.L. (1972): J. Chem. Phys. **56** 1652
Hermann, D.S., Rhodin, T.N. (1966): J. Appl. Phys. **37** 1594
Jaycock, M.J., Parfitt, G.D. (1981): "Chemistry of Interfaces" (Halsted Press: a Division of John Wiley & Sons, New York e.a.)
Jonsson, H., Weare, J.H., Levi, A.C. (1984): Surf. Sci. **148** 126
Kadanoff, L.P., Baym, G. (1962): "Quantum Statistical Mechanics. Green Function Method in Theory of Equilibrium and Nonequilibrium Processes" (W.A. Benjamin, Inc., New York)
Kashchiev, D. (1976): Surf. Sci. **55** 477
Kashchiev, D. (1977): J. Cryst. Growth **40** 29
Keizer, J. (1987): "Statistical Thermodynamics of Nonequilibrium Processes" (Springer Verlag, New York e.a.)
Kern, R., Le Lay, G., Metois, J.J. (1987): Curr. Top. Mater. Sci. **3** 139
Kolesnichenko, E.G. (1986): "Nonequilibrium Effects in Dynamics of Chemically Reacting Gaseous Mixtures", in "Physical - Chemical Kinetics in Gas Dynamics" (Moscow State Univ., Moscow), pp. 80–100 (in Russian)
Koropov, A.V., Sagalovich, V.V. (1990): Poverkhnost (USSR) No. 2 17 (in Russian)
Kreuzer, H.J. (1990): Surf. Sci. **231** 213
Kreuser, H.J., Payne, S.H. (1988a): Surf. Sci. **200** L443; (1988b): Surf.Sci. **205** 153
Kreuser, H.J., Payne, S.H. (1989): Surf. Sci. **222** 404
Kudryavtsev, I.K. (1987): "Chemical Instabilities" (Moscow State Univ., Moscow) (in Russian)
Landau, L.D., Lifshitz, E.M. (1965): "Quantum Mechanics. Non-Relativistic Theory", 2nd ed. (Addison-Wesley, New York)
Landau, L.D., Teller, E. (1936): Phys. Z. Sowjet **10** 34
Lewis, B., Anderson, J.C. (1978): "Nucleation and Growth of Thin Films" (Academic Press, New York)
Lipkin, H.J. (1973): "Quantum Mechanics" (North-Holland, Amsterdam)
Lundquist, B.J., Gunnarson, O., Hjelmberg, H., Nørskov, J.K. (1979): Surf.Sci. **89** 196
Mantell D.A., Ryali, S.B., Haller, G.L., Fenn, J.B. (1983): J. Chem. Phys. **78** 4250
Manson, J.R. (1991): Phys. Rev. B **43** 6924
Mayorov, A.A., Filaretov, A.G., Tzyrlin G.E. (1988): "Statistical Model of GaAs Growth in Molecular Beam Epythaxy Method", in: "Scientific Instrumentation" (USSR) (Nauka, Leningrad) pp. 98–103
Morris, W.L., Hines, R.L. (1970): J. Appl. Phys. **41** 2231
Murch, G.E., Thorn, R.J. (1979): Phil. Mag. A **40** 477
Moses, C., Zeng, Hong, Lin. J.S., Li, Wei, Karimi, M., Vidali, G. (1992): J. Vac. Sci. Technol. A **10** 2377
Nayfeh, A.H. (1981): "Introduction to Perturbation Techniques" (John Wiley and Sons, New York)
Newns, D.M. (1985): Surf. Sci. **154** 658
Niven, W.D., Ed. (1972): "The Scientific Papers of James Clerk Maxwell" (Dover, New York), Vol. 2
Nørskov, J.K., Stoltze, P. (1987): Surf.Sci. **189/190** 91
Nourtier, A. (1985): J. de Phys. **46** 55
Osipov, A.V. (1990): "Kinetics of Nucleation of Thin Films on Solid Surfaces from Gas Phase": Ph. D. Thesis (St.Petersburg) (in Russian)
Pechukas P., Davis, J.P. (1972): J. Chem. Phys. **56** 4970
Pereira, V., Zgrablich, G. (1989): Surf. Sci. **209** 512
Popov, V.N. (1983): "Functional Integrals in Quantum Field Theory and Statistical Physics" (Reidel, Boston)

Rajaraman, R. (1982): "Solitons and Instantons" (North-Holland, Amsterdam)
Reed, D.A., Ehrlich, G. (1981): Surf. Sci. **216** 23
Ruzaykin, M.R., Ervye, Yu.Yu. (1989): Poverkhnost (USSR) No. 4 5 (in Russian)
Sacedon, J.L., Martin, C.S. (1972): Thin Solid Films **10** 99
Scott, C.D. (1980): "Catalytic Recombination of Nitrogen and Oxygen on High Temperature Reusable Surface Insulation", AIAA Pap. No. 1477, pp. 1–9
Sergeev, V.L. (1978): Vestnik LGU (USSR) No. 13 98 (in Russian)
Tiganov, I.I. (1987): "Analytical Parametrizations of Kinetic Coefficients in Real Gas Dynamics": Ph.D.Thesis (St.Petersburg) (in Russian)
Trofimov, V.I. (1975): Fiz. Tv. Tela (USSR) **17** 2478 (in Russian)
Venables, J.A. (1973): Phil. Mag. **27** 697
Venables, J.A., Spiller, C.D., Nanbücken, M. (1984): Rep. Prog. Phys. **47** 399
Voloshchuk, V.M. (1984): "Kinetic Theory of Coagulation" (Gidrometeoizdat, Leningrad) (in Russian)
Voter, A.F., Doll, J.D. (1984): J.Chem.Phys. **80** 5832
Wetter, K. (1967): "Electrochemical Kinetics" (Khimiya, Moscow) (in Russian)
Zallen, R. (1988): "The Physics of Amorphous Solids" (Academic Press, New York)
Zandberg, E.Ya., Ionov, N.N. (1969): "Surface Ionization" (Nauka, Moscow) (in Russian)
Zener, C. (1931): Phys. Rev. **38** 277
Zinsmeister, G. (1968): Thin Solid Films **2** 497; (1969): Ibid **4** 363; (1971): Ibid **1** 51
Zhdanov, V.P. (1985): Fiz. Tv. Tela (USSR) **27** 2573 (in Russian)
Zhdanov, V.P. (1988): "Elementary Physical - Chemical Processes at Surface", (Nauka, Novosibirsk) (in Russian)
Zgrablich, G., Pereira, V., Ponzi, M., Marchese, J. (1986): American Institute of Chemical Engineers Journal **32** 1158
Zubcus, V.E., Tornau, E.E. (1989): Surf. Sci. **216** 23

Springer-Verlag and the Environment

We at Springer-Verlag firmly believe that an international science publisher has a special obligation to the environment, and our corporate policies consistently reflect this conviction.

We also expect our business partners – paper mills, printers, packaging manufacturers, etc. – to commit themselves to using environmentally friendly materials and production processes.

The paper in this book is made from low- or no-chlorine pulp and is acid free, in conformance with international standards for paper permanency.